T0353568

AI for Behavioural Science

This book is a concise introduction to emerging concepts and ideas found at the intersection of contemporary behavioural science and artificial intelligence. The book explores how these disciplines interact, change, and adapt to one another and what the implications of such an interaction are for practice and society.

AI for Behavioural Science begins by exploring the field of machine behaviour, which advocates using behavioural science to investigate artificial intelligence. This perspective is built upon to develop a framework of terminology that treats humans and machines as comparable entities possessing their own motive power. From here, the notion of artificial intelligence systems becoming choice architects is explored through a series of reconceptualisations. The architecting of choices is reconceptualised as a process of selection from a set of choice architectural designs, while human behaviour is reconceptualised in terms of probabilistic outcomes. The material difference between the so-called "manual nudging" and "automatic nudging" (or hypernudging) is then explored. The book concludes with a discussion of who is responsible for autonomous choice architects.

Stuart Mills is a behavioural economist with a background in economics and political economy. His research focuses on nudge theory, personalisation, and digital economy. He is interested in the intersection of technology, data, and behavioural science within public policy and finance, as well as the wider political economy implications.

AI for Behavioural Science

Science

Stuart Mills

CRC Press
Taylor & Francis Group
Boca Raton London New York

CRC Press is an imprint of the
Taylor & Francis Group, an **informa** business

First Edition published 2023
by CRC Press
6000 Broken Sound Parkway NW, Suite 300, Boca Raton, FL 33487-2742

and by CRC Press
4 Park Square, Milton Park, Abingdon, Oxon, OX14 4RN

CRC Press is an imprint of Taylor & Francis Group, LLC

© 2023 Stuart Mills

ISBN: 978-1-032-06669-1 (hbk)
ISBN: 978-1-032-06401-7 (pbk)
ISBN: 978-1-003-20331-5 (ebk)

DOI: 10.1201/9781003203315

Typeset in Times
by Deanta Global Publishing Services, Chennai, India

Contents

Introduction

This is a brief introduction. I mean this in two ways. Firstly, in the most literal sense – I intend this introductory chapter to be short. Secondly, there are many discussions which could fall under the umbrella of "artificial intelligence for behavioural science," and this book could not possibly cover them all. Therefore, I have restricted my focus within this book to how artificial intelligence (AI) systems can be used to influence individual behaviours through the altering of decisional-contexts (or *choice architecture*).

But I have also tried to write a book that gives a reader tools to begin their own inquiry into this vast and rapidly developing area, which one might broadly call *behavioural technology*. For behavioural scientists, I offer various models, concepts, and I hope useful definitions, which will allow one to understand how behavioural science is *translated* to a machine. For computer scientists, I offer reflections on how such translation can diminish or perhaps distort the human subject at the centre of all these shenanigans.

Thematically, various dualities appear in this book, for instance, human *vs.* machine, individual *vs.* population, potency *vs.* utility, understanding *vs.* convenience, computational logic *vs.* messy realities, labs rats *vs.* fellow test subjects. I do not focus explicitly on any of these dualities, but they are characters throughout this book. This is inevitable. *Artificial intelligence for behavioural science* is a collision of two disciplines, and prominent, accepted ideas in one discipline may appear strange, or be wholly *absent*, for another. This should provide a clue as to what I mean by *translation* – the process of allowing these dualities to play out and seeing what happens.

An example comes when considering the subject this book is about. I assert throughout this book that *artificial intelligence for behavioural science* is qualitatively different from *behavioural science for artificial intelligence*. As I will discuss in Chapter 1, *behavioural science for artificial intelligence* represents the *use of behavioural science* within *artificial intelligence* development. This is not the subject which this book is about; this book is about how AI can be used within behavioural science. While the alternative is quite interesting, it would be a mistake to interpret these perspectives as one and the same.

In some ways, the fields of artificial intelligence (AI) and behavioural science (insofar as it is focused on the human subject) are two sides of the same

DOI: 10.1201/9781003203315-1

coin. While AI researchers endeavour to "create mind in machines" (Turkle, 1988, p. 244), the objective of behavioural science (and the fields which constitute it, namely, psychology and – to a lesser extent – economics) is to explain the product of the mind (i.e., behaviour) within the world (Haig, 2014). The similarities between fields, therefore, originate from the central objective of probing the "black box" of intellect and reason (and thus *behaviour*) which is possessed by the human, and is given to the machine. Naturally, the history of AI and behavioural science is one which is entangled (Miller, 2003; Turkle, 1988).

Yet, these disciplines are not the *same*; they are not *co-dependents* on some intellectual journey. Therefore, I prefer to talk about the *co-history* of AI and behavioural science. By a co-history, I mean two histories which can interact with and influence one another but can also be viewed quite independently. I want this discussion – and this book – to do justice to this character and to not falsely equate one with the other. Just because, on some level, both AI and behavioural science investigate the same entity (i.e., mind) or phenomena (e.g., behaviour, decision-making, reasoning, intelligence) does not mean that a development in, say, AI is born from or results in a development in behavioural science, and *vice versa*.

A simplified co-history begins with economics. Economics is a field concerned with optimisation, and this extends to questions about human behaviour. Von Neumann and Morgenstern's (1944) *Theory of Games and Economic Behaviour* is possibly the most famous attempt to *axiomatically* describe rational human behaviour (i.e., the "rules" of rationality).[1] Such axioms proposed a model of human behaviour where a person would, for instance, be consistent in their preferences, or would maximise their expected utility. For AI research, such a model was a worthwhile launching pad for thinking about how to make a machine intelligent (Russell, 1997), though this approach was soon critiqued and revised following powerful interventions by thinkers such as Herbert Simon (1955, 1956, 1994 [1969]), supported in part by Miller's (1956) work on human cognitive limitations (see Chapter 2). Simon's notion of *bounded rationality* would later form the foundation of the critique of the axiomatic approach as a *description* of human behaviour given by Tversky and Kahneman (1973, 1974, 1979) within cognitive psychology.[2] From such critiques, the disciplines of behavioural economics and behavioural science would emerge (Thaler, 2015).

Bounded rationality as an idea also had a profound impact on what AI was expected to be (Russell, 1997). Accepting that computers were bounded in their computational power, just as humans were in their cognitive power, was an important idea which shaped the development of early "symbolic" AI.[3] Symbolic AI systems follow logical rules which allow one to determine what action one should take, given some input. This potentially computationally

taxing approach spurred much of Simon's research into human and organ-
isational decision processes as a means of understanding how these intelligent entities – possessive of less computational power than a computer – made
intelligent decisions (Simon, 1994 [1969]).

Nevertheless, progress in symbolic AI slowed, while computational power
grew exponentially, as did available data. By the 1980s, nodal-linked AI was
coming to dominate AI research (Turkle, 1988). Nodal-linked AI, more commonly known today as *neural networks* and *deep learning*, are systems loosely
inspired by the structure of the human brain and built from machines called
perceptrons (Minsky and Papert, 2017 [1969]). Perceptrons can take in large
amounts of data, "combine" these data using weights, and make a prediction
based on this computation (e.g., cat or dog). Then, the prediction is compared
with a known answer (e.g., Prediction: 87% cat; Answer: 100% cat), and the
weights used to "combine" the data are adjusted so that the next prediction is
closer to the known answer. This approach builds off ideas of connectionism
in psychology (Hayek, 1999 [1952]) and earlier understandings of neurons in
the human brain (von Neumann, 2000 [1958]), whereby connections become
stronger through practice and repeat exposure to an input stimulus.[4] Nodal-
linked AI is today the dominant approach to AI, and it is the approach primarily drawn upon throughout this book (see Chapter 1).

During the period that nodal-linked AI was developing, the "heuristics
and biases" programme which Tversky and Kahneman (1974) had outlined
was expanding and developing into behavioural economics and what I will call
throughout this book *contemporary* behavioural science. Before this revolution
in psychology, the discipline had been dominated by *behaviourism*, a school of
psychological thinking which contends internal mental states are unnecessary
to understand behaviour. Instead, one can understand human behaviour by
having a good (by which one means *probabilistic*; Skinner, 1984) understanding of what stimulus will produce a desired behaviour (see Chapter 5). The
cognitive psychology approach, of which Tversky and Kahneman were a part,
which followed behaviourism took no less of a view than the behaviourists that
the mind was a "black box" (Haig, 2014; Skinner, 1976 [1974]; Turkle, 1988),
but contended that understanding the black box was also key to understanding
human behaviour.[5]

By the 1990s, studies into human bias and decision-making were producing an impressive body of work which professed to provide a deeper understanding of human behaviour than models such as economics' rational model
ever could. *Anomalies*, written by Kahneman, Knetsch, and Thaler (1991),
would stand as an important attempt to combine key ideas from this intellectual revolution, including loss aversion (Kahneman and Tversky, 1979), the
endowment effect (Knetsch, 1989), and the status quo bias (Samuelson and
Zeckhauser, 1988; see Chapter 3). Further development of these ideas would

occur throughout the 1990s,[6] resulting in various investigations into the *application* of these theories of human behaviour to *change* behaviour (e.g., Johnson and Goldstein, 2003; Madrian and Shea, 2001; Thaler and Sunstein, 2003). These efforts culminated in Thaler and Sunstein's (2008) book *Nudge*, which suggested that the context in which humans make decisions (so-called *choice architecture*) could be actively redesigned, drawing on prior behavioural theories, to positively influence the behaviour of decision-makers. Contemporary behavioural science had arrived.

*

I am now at the point where it seems appropriate to begin this book. The co-history I have given is a snapshot, not a complete account. But a reader of this book, I *presume*, is far more interested in reading about how the disciplines of artificial intelligence and behavioural science are colliding once more than how they have collided in the past. Though, as the reader will also come to see, such background is necessary to understand where (I think) this story is going next. Chapter 1 considers *behavioural science for artificial intelligence*, while the remainder of this book focuses on *artificial intelligence for behavioural science*. Chapter 2 introduces some useful definitions which will establish the language framework upon which the rest of this book is built. Chapter 3 introduces the concept of the *autonomous choice architect* and offers several reconceptualisations of behavioural science and *nudging* such that they may be understood and performed by, respectively, an AI system. Chapter 4 introduces the concept of the *hypernudge* and explores the consequences of individuals interacting with highly efficient nudging machines. Finally, Chapter 5 turns to the ethical implications by asking who is responsible for AI in behavioural science, as well as questioning some of the assumptions and logics which arise from using AI in behavioural science.

NOTES

1. An earlier work often offered in such discussions is Mill's (1836) essay *On the Definition of Political Economy and of the Method of Investigation Proper to it*, though in my opinion this is something of a simplification of the argument Mill makes. Prior to von Neumann and Morgenstern's (1944) work, where economics was less mathematical and more resemblant of political economy, musings on human behaviour as an irrational agent were quite common. For instance, see Keynes' (2017 [1936]) various theories, including *liquidity preference, animal spirits,* and *behaviour under uncertainty*, which all retrospectively could be understood as a kind of behavioural economics (Thaler, 2015).

2. As Heukelom (2012) notes, Tversky and Kahneman do not seem to have been aware of Herbert Simon or bounded rationality at the time they were developing their "heuristics and biases" programme, and indeed, the resurrection (in a manner of speaking) of Simon's notion within behavioural science has been one in support of, and wrapped in the language of, the heuristics and biases programme (Gigerenzer, 2007). As Heukelom (2012) has argued, this does leave some important differences between these perspectives to be missed, notably that Simon's interest concerns *how people decide*, whereas Tversky and Kahneman's interest concerns *how people simplify to be able to decide*.

3. Symbolic AI, or as referred to throughout this book, rule-based AI, is AI that is designed to follow various logical rules depending on the input it is given. For instance: IF X = 1, choose A; ELSE, choose B. All rules can be described in terms of a *system of symbols* (e.g., mathematical symbols, logical instruments, musical notation), hence the notion of AI being symbolic. Symbolic AI has some advantages. For instance, it is relatively easy to understand what the AI is doing at any given moment. It also has several disadvantages, primarily that not every task is easily described symbolically. Minsky and Papert (2017 [1969]) illustrate this with what one could call the cat problem:

> Using any set of symbols or semantics, and as many rules or logical statements as one wishes, derive a set of statements which accurately identify a cat, and which do not identify any non-cat entity as a cat.

With enough time, and enough rules (and perhaps enough knowledge of cats), one may be able to accomplish the task. But as the set of statements must be so specific, such a task is likely impossible for any intelligent entity to complete. Specificity is a primary cause of this difficulty. As all cats have some variation to them, statements which are so specific as to exclude all non-cats must also be flexible enough to allow for variation within all cats. Achieving such a balance is difficult. Perceptron-based AI (what I refer to as nodal-linked AI) resolves this problem by never actually articulating *how it determines if something is a cat*.

4. Neural networks consist of several (many) perceptrons, while deep learning describes a neural network with several (many) hidden "layers" – essentially, a big neural network. These AI systems are currently the most successful in terms of tasks such as identifying cats from dogs or beating humans at games such as *Jeopardy* and *Go*. Quite how these systems determine whether something is, say, a cat rather than a dog (which is to say, why the weights are set as they are) remains an outstanding question. In this sense, knowledge is said to be *entangled* within a neural network. In symbolic AI, one need only know the rule, and the current input, to be able to understand the whole process. Symbolic AI is said to be *disentangled*. Emerging ideas regarding *neurosymbolic AI* seek to combine both approaches to resolve challenges which are representative in one paradigm, but not the other. See, for instance, Garcez and Lamb (2020) and Garnelo and Shanahan (2019).

5. Following Turkle's (1988) account, part of what "killed" behaviourism was the invention of the computer. The argument goes that the computer, on the face of it, was a behaviourist machine, insofar as one could – just as with humans and animals – give the computer inputs (stimuli) and observe outputs (behaviours).

But unlike humans and animals, computer scientists could tinker with the internal structure of the computer, and this would change the relationship between inputs and outputs. Turkle (1988) argues that the computer demonstrated the importance of mental states in intelligent entities, and therefore struck a deathblow to behaviourism. This is likely too simple an explanation. *Firstly*, thinkers such as Chomsky (1959) are often held to have contributed to the decline of behaviourism. Chomsky's (1959) theory of universal grammar suggested there may be innate properties to human behaviour and intelligence, which contradicted behaviourist ideas about reinforcement learning (see Chapter 5). *Secondly*, behaviourist approaches were better suited for relatively simple, controlled environments, and as key areas where behaviourism was applied became more complex (e.g., manufacturing), behaviourist theories failed to provide a satisfactory approach.

6. For instance, Tversky and Kahneman (1992) would revise and update their original prospect theory, which derived loss aversion, while Benartzi and Thaler (1995) would incorporate a temporal dimension.

Human Behaviour and Machine Behaviour

1

MACHINE BEHAVIOUR

This book proceeds on the basis that artificial intelligence for behavioural science is qualitatively different from behavioural science for artificial intelligence. The bulk of this book, therefore, is orientated towards the former. Nevertheless, it is worth taking some time to consider the latter.

What I have described here as *behavioural science for artificial intelligence* has been dubbed by Rahwan *et al.* (2019, p. 477) "machine behaviour," which they define as "the scientific study of behaviour exhibited by intelligent machines" (pp. 477–478). Pedersen and Johansen (2020, p. 519, original emphasis) also offer the term "artificial intelligent behaviour" and "argue that due to the complexity and opacity of *artificial inference*, one needs to initiate systematic empirical studies of artificial intelligent behaviour similar to what has previously been done to study human cognition, judgment and decision making." I will adopt the language offered by Rahwan *et al.* (2019) henceforth.

Understanding Black Boxes

Artificial Intelligence (AI) is increasingly seen throughout society, being used to curate news items, provide credit scores, set product prices, trade financial products, direct traffic around cities, manage the state of our homes, match potential romantic partners, determine educational outcomes, and determine targets in armed conflicts. These are all socially important functions which demand accountability and oversight for the systems engaging in them (Floridi,

DOI: 10.1201/9781003203315-2

2020; Pedersen and Johansen, 2020; Zerilli *et al.*, 2019). Yet, AI systems often suffer from a lack of transparency, and can be tremendously opaque, even to those experts who programme them (Burrell, 2016). They are black boxes (Pasquale, 2015; Turkle, 1988).

Rahwan *et al.* (2019) offer four reasons for the lack of transparency in AI functionality, which Pedersen and Johansen (2020) and Burrell (2016) broadly echo:

1. AI systems are computationally complex, meaning it can be difficult and time consuming to unpick exactly what the system is doing, even for their human programmers. Furthermore, AI can vary dramatically depending on the task which it is programmed to perform. Therefore, even an expert in AI may struggle to understand different AI systems because different systems are designed to accomplish different objectives.
2. AI systems and their input data are often proprietary, and so understanding a system requires either inside knowledge of the AI design and input data or the systematic testing of different inputs and observation of corresponding outputs. In the first instance, those who have this access are likely human programmers who suffer from point (1). In the second instance, understanding can only ever be based on inference.[1]
3. AI systems can change and learn, meaning even if one can fully grasp how a system is programmed, and even if one has a good inferential theory of the various inputs and outputs, these are likely to change over time. This is particularly disruptive for those building inferential theories, as the process of experimenting with different inputs may itself lead to change in the AI system.
4. AI systems do not operate in isolation but in fact interact with other AIs (as is frequently the case with algorithmic trading in financial markets) and human beings. Such systems are *hybrid* systems.[2] Because outside influences may be constantly interacting with an AI system, causing it to change, an experimenter may suffer the same issue as in point (3). Equally, because an AI programmer can never predict how outside influences, such as people, will interact with the AI system, even an expert may not know how the AI will change and develop.[3]

If AI systems are increasingly important to society, and increasingly demand accountability, understandability, and oversight so as to not disrupt society unduly but are simultaneously opaque and difficult to understand or explain, how might one proceed? This is, broadly, the problem statement which motivates the machine behaviour programme.[4] While Rahwan *et al.* (2019) argue for a broad integration of computer science and behavioural science to make progress in this domain, it is Pedersen and Johansen (2020) who offer a more intuitive programme for proceeding.

As alluded to above, one method of understanding AI may be through a process of *inference*, and it is this process which Pedersen and Johansen (2020) advocate. They argue that just because a phenomenon cannot be directly observed does not mean insight cannot be gained. They support this argument with reference to the human mind. The human mind is, in several ways, as much of a black box as an AI "mind" (Ashby, 1978; Simon, 1981, 1977; Turkle, 1988). Yet contemporary behavioural science (e.g., Kahneman, 2011, 2003) can produce theories of human cognitive function not by observing the human mind directly (e.g., consciousness, subconsciousness) but by observing human behaviour in response to various stimuli (Friesen, 2010; Miller, 2003). Pedersen and Johansen (2020) argue that such observations allow behavioural scientists to infer human cognitive processes and that through similar techniques, one can infer the inner workings of an AI system (Haig, 2014).

For instance, imagine an AI system in which, when the number 2 is entered, the number 5 is returned. There are multiple ways in which this input, 2, can be transformed into the output, 5, such as doubling the input and adding 1 (i.e., $f(x) = 2x + 1$), or simply adding 3 to the input (i.e., $f(x) = x + 3$). A machine behaviour approach would then suggest choosing another input, say 4, and observing the output. If the AI is doubling and adding 1, an output of 9 would be predicted. If it is adding 3 to the input, an output of 7 would be predicted. This approach represents *abductive* reasoning (Pedersen and Johansen, 2020) insofar as there is continuous experimentation with inputs and outputs to discern a more reliable theory of the process. Note, of course, that even if the output is 7 when the input is 4, one does not *know* the AI is just adding 3 (Forde and Paganini, 2019; Lipton and Steinhardt, 2018). There could be a wholly different process which accounts for these two observations.[5] Nevertheless, Pedersen and Johansen (2020) assert that this process of studying *artificial behaviour* can shed light on the behaviours of machines, including whether a machine is biased.

A reflection on some key literature within this intellectual space aids understanding of Pedersen and Johansen's (2020) perspective.

Consider two examples:

1. *The Status Quo Bias*, which is closely related to the default option effect and decisional inertia, describes the tendency for people to stick with whatever choices/circumstances are already determined (Samuelson and Zeckhauser, 1988). Several explanations of this tendency, or *bias*, have been offered, including the idea that the status

quo can be seen as an endorsement or implicit recommendation (Jachimowicz *et al.*, 2019; Madrian and Shea, 2001; McKenzie *et al.*, 2006; Thaler and Sunstein, 2008), that people want to avoid the risk of losing out by switching (Jachimowicz *et al.*, 2019; Kahneman and Tversky, 1979), and simply that it is easier to stick with the status quo than to change (Johnson *et al.*, 2012).

2. *The Social Norm Bias* (e.g., conformity) describes the tendency for people to "follow the path most-travelled," which is to say, to do as others do (Bernheim, 1994). Several explanations for this tendency, or *bias*, have also been offered, including the idea that conformity enables one to demonstrate their standing in society (Bernheim, 1994; Veblen, 2012 [1899]), that conformity is motivated by a desire to avoid social shunning and shame (Sunstein, 1996), and that following others allows one to borrow knowledge, or supplement their own ignorance (Moon, 2010).

In both examples, there is ample evidence to support the "outputs" of status quo bias or social norm bias, but explanations of how the "inputs" (i.e., stimuli) and "outputs" (i.e., behaviour) relate have resulted in a multitude of theories, just as the inferential method offered by Pedersen and Johansen (2020) does not determine what the system is doing but allows one to narrow the set of theories which the system *could* be following (Haig, 2014).[6] This is, however, a partial misrepresentation of Pedersen and Johansen (2020). By using an example (i.e., the number-transformer AI) which ignores the possibility that the AI system changes as different inputs are tested and outputs observed (i.e., point (3), above), there is an over-emphasis placed on the notion of falsification (Popper, 1959), something which is generally not found in contemporary behavioural science.[7] By accepting AI systems can and do change, thus making falsification a much more difficult standard to adhere to, the programme of machine behaviour Pedersen and Johansen (2020) advocate to investigate AI "black boxes" does seem to align reasonably with the programme contemporary behavioural science uses to investigate human "black boxes" (i.e., the mind; Haig, 2014).

Are Machines Behaving?

The key question at this stage is this: is the process of transforming an input into an output something which tells an observer about the *behaviour* of the machine? This is not a question which is easy to answer, and it benefits from a few additional examples.[8] One immediate idea to draw upon comes from Rahwan *et al.* (2019), who argue, in addition to that discussed by Pedersen and Johansen (2020), that one must consider various practical aspects of the machine's design. For instance, imagine two computers, A and B, both of which contain the above number-transformer programme. Perhaps the number-transformer process is computationally intensive, and computer A is an outdated laptop while computer B is a state-of-the-art supercomputer. Computer A may take hours or days to return a result or may simply give an error message. Computer B may be nearly instantaneous. Is it accurate to describe these computers as *behaving* differently?[9]

Consider a second example. *Word2Vec* is a computational process for vectorising individual words so as to allow mathematical computation on natural language (Mikolov *et al.*, 2013).[10] This is done using a neural network which predicts word associations.[11] As such, it is possible to give *Word2Vec* questions such as "King is to Man as Queen is to what?" and the programme will, often, return the word "Woman." As this example question suggests, *Word2Vec* can "learn" various word associations. The programme may also appear biased. For instance, if the statement, "Man is to Strong as Woman is to what?" is given to the programme, and the programme returns the word "Weak," this is clearly a sexist word association, and a machine behaviourist may determine the programme is sexist. Yet is a programme which has learnt natural word associations from natural language text given to it by a programmer *actually* sexist, or does the programme's response indicate that *natural language* often contains sexist word associations?

The machine behaviour approach described here often assumes a degree of independence between the AI system and the programmer, the training data, and the user. Indeed, without this assumption, machine behaviour could simply be seen as the study of the behaviour of the programmer, or the mass behaviour of those within a dataset, via a computer-intermediary. The human aspect of machine behaviour is acknowledged by both Rahwan *et al.* (2019) and Pedersen and Johansen (2020). Rahwan *et al.* (2019, p. 480) write:

> In the context of machines, we can ask how machines acquire (develop) a specific individual or collective behaviour. Behavioural development could be directly attributable to human engineering or design choices. Architectural design choices made by the programmer (for example, the value of a learning rate parameter, the acquisition of the representation of knowledge and state, or a particular wiring of a convolutional neural network) determine or influence the kinds of behaviours that the algorithm exhibits ... A human engineer may also shape the behaviour of the machine by exposing it to particular stimuli. For instance, many image and text classification algorithms are trained to optimize accuracy on a specific set of datasets that were manually labelled by humans. The choice of dataset – and those features it represents – can substantially influence the behaviour exhibited by the algorithm.

From the above, perhaps, then, machine behaviour really is the study of behaviour demonstrated via machines but originating from humans. After all, a common argument within the algorithmic bias literature is not that the *algorithm* is biased but the data fed to the algorithm is biased, either because those who generate the data *exhibit* bias, or those who *select* the data are biased (Buolamwini and Gebru, 2018). Yet, Pedersen and Johansen (2020, p. 526) continue to distinguish between "the transfer of biases from programmers," in effect acknowledging the role of programmer bias, and "machine-generated biases." But what are machine-generated biases? Despite great efforts by the respective authors, an immediate answer does not seem forthcoming.

One explanation may be that these authors do not mean to understand machine behaviour from a single AI system (Pedersen and Johansen, 2020; Possati, 2020). It is perhaps right to be sceptical of labelling *Word2Vec* biased when only looking at a single neural network model, as the influence of the programmer in terms of programming choices, hardware selections, and data selections are much greater here. Pedersen and Johansen (2020) instead argue that machine behaviour should seek to study multiple AI systems, engaged in a plethora of tasks, in a systematic fashion. In this way, the individual choices of programmers are diminished, while – presumably – a wider range of data sources, choices, and uses are examined. Rahwan *et al.* (2019) offer a similar perspective, going so far as to describe the various system preferences for AI programmers as representing a kind of computational evolution. This approach, while admirable, may still face two challenges.

The first relates to point (1) offered above. Where AI systems differ, it may be difficult to draw valid conclusions about behaviour. In behavioural science, humans often differ greatly (Bryon, Tipton, and Yeager, 2021; Mills, 2022). Yet, human beings can switch from one simple task (say, adding up two two-digit numbers) to another simple task (say, recalling what they had for breakfast) with relative ease. This has allowed behavioural scientists to study human cognitive processing in a systematic fashion, being able to identify and control for various differences. By contrast, an AI which is designed to trade stocks is not easily comparable to an AI which has been tasked with managing taxi availability in a city.[12] Of course, there are points of comparison: the first AI likely has assigned some value to stock A and based on this value chooses whether to buy or sell, while the second AI likely has assigned some value to taxi B and based on this value chooses whether to move the taxi or keep it where it is. Machine behaviourists may be able to compare the assignment function and decision function of each of these AIs and gleam insights. Yet, as an astute reader will note, this apparent comparability is, in large part, possible because of human (i.e., my) rationalisation of AI processes. In comparing these two AIs, a machine behaviourist must *first decide these functions exist*, and only *then* go looking for them.[13] As Possati (2020, p. 10) writes, "what makes a set of circuits and logic an AI system is the human interpretation of this set. What makes a function a computation is not the function itself, but the human interpretation of this function."[14]

The second relates to the above argument that much of the behaviour attributable to machines actually reflects human behaviour and human bias. Rahwan *et al.* (2019) make this quite apparent insofar as they talk about the evolution of AI systems. They argue that rather than AI systems actually evolving through the process of mating and random mutation, AI systems are better described as *developing*, and developing within institutional settings, with the various habits, norms, and limitations of the programmer feeding into the AI design. In the evolutionary sense, Rahwan *et al.* (2019, p. 481) argue, this can create *path dependence*:

> At each step, aspects of the algorithms are reused in new contexts, both constraining future behaviour and making possible additional innovations. For example, early choices about microprocessor design continue to influence modern computing, and traditions ... build in many assumptions and guide future innovations by making some new algorithms easier to access than others.

This is an interesting perspective, and to an extent, the question of technology shaping behavioural possibilities (Frischmann and Selinger, 2018) is an idea to which I will return later in this book. It is also, to an extent, a question of philosophy of science insofar as one wonders why researchers and developers choose, within one paradigm, to focus on a particular approach, and within another, to focus on a different – perhaps radically different – approach (Kuhn, 2012 [1962]).[15] Yet, despite the interesting questions this perspective raises, it seems to do little to explain what "machine-generated biases" are or to inform the present inquiry into whether AI systems can be biased or can exhibit their own behaviour. Any behaviours and biases which emerge via this route seem firmly human in origin, even if the human origin is dispersed across time and institutions.

Machines as Mirrors

In Chapter 2, I will explore terms such as "behaviour" in more depth. But presently, the question of machine behaviour seems to have reached something of an impasse. Machines – at the time of writing – do as they are told,[16] and if one is interested in what a machine does (i.e., machine behaviour), then it seems the place to begin is with the person telling the machine what to do. Furthermore, AI systems learn from the data they are given, which reflects two human tasks, *generating* data and *selecting* data. In the case of *Word2Vec*, for example, it seems inaccurate to say the programme is biased, when the programme is merely identifying word associations which humans may make in natural conversation. Likewise, it seems silly to contend that AI systems *behave differently* because humans have chosen to run the same programme on different hardware.

To an extent, this proclivity to anthropomorphise machines and "their" actions, behaviours, and identities is quite a natural one. For instance, popular machine learning AI systems, such as neural networks, follow connectionist theories[17] of the human mind (Minsky and Papert, 2017 [1969]; Watson, 2019), while Turkle's (2004 [1984]) authoritative study of early human–computer interactions frequently draws on the idea of computers in general having human-like characteristics (Ashby, 1978; Turkle, 1988; von Neumann, 2000 [1958]). The reverse is also increasing true (Sætra, 2021b). Simon's (1994 [1969], 1981) early ideas about rationality were informed by reflections about human "processing power" compared to a computer's, and it is now common parlance to talk of "hacking" one's life. Indeed, there is a whole school of philosophy known as *posthumanism* which seeks to draw commonalities between humans and non-humans and opens up the way for discussions of humans and computers perhaps representing similar, if not *interchangeable*, entities (Coeckelbergh, 2020). With all of this in mind, it does not seem so great a sin to (a) talk of machines as *behaving* and (b) attribute human qualities to machines. It is worth taking some time to explore some of the avenues of thought unlocked from this perspective.

Returning to Turkle (2004 [1984]), the story is somewhat more complex than computers simply being attributed to human characteristics. Turkle's (2004 [1984]) study is not about computers but rather the people who use them. Her conclusions impact the subject: "Whether or not computer scientists ever create an artificial intelligence that can think like a person, computers change the way people think – especially about themselves" (Turkle, 2004 [1984], p. 152).[18] Possati (2020) offers a comparable account, building on the machine behaviour perspective. Possati (2020) argues that a psychoanalytic perspective can enhance the study of machine behaviour by directing investigators to not just study the behaviour of AI systems but to see AI systems as artefacts of human behaviours and identities.[19] In short, computers can act as mirrors for our own humanity and human traits. This discussion greatly benefits from the exercise of seeing computers and humans, at least in *some capacity*, as being comparable entities.[20]

This perspective offers something of a reprieve from the criticism of machine behaviour I have presented thus far. Perhaps AI systems reflect the biases and behaviours of those humans who create them; but in studying AI systems in the way described by machine behaviour, one can begin to see the biases and behaviours of humans which were perhaps previously masked.

Consider once more *Word2Vec*. No human could ever identify the various word associations which may exist throughout Wikipedia or *The New York Times*, for instance. But just because a human cannot identify them does not mean associations (good or bad) are not being made. By employing a programme like *Word2Vec*, these associations are revealed, and humans can see how natural language reflects various attitudes. *The machine acts as a mirror.*

Programming Conflict

Another interesting perspective emerges from considering humans and computers as equivalent entities. I have argued it is unfair to call AI systems and associated algorithms biased on the basis that bias seems to always originate from the humans behind the AI system, either directly (i.e., the programmer) or indirectly (e.g., via input data, via human interaction). But if one draws an equivalence, might it not be tempting to argue that *humans cannot be biased either*? For instance, if a person tends to talk in terms of *mankind*, and associates *male* pronouns with ideas such as strength, leadership, and fortitude, is that person biased in their language? Or, rather, is it that they have *learnt* those stereotypes from a larger social structure, say a patriarchal social structure which leads these associations to exist implicitly within the mind of this hypothetical subject? This is a big question, and neither I nor this book is equipped to offer a satisfying answer. The purpose of asking this question, however, is to reveal where humans and machines may differ.

Imagine a person and an AI exhibit the same word associations which appear to demonstrate gender bias. If the AI is told this, it will probably not understand. In the first instance, it must be equipped with a *means to be told*. Assuming it is, it must then be equipped *with a means of doing something about it*. The latter is surely more difficult than the former. Indeed, if the computer could speak, one suspects it would throw an accusation straight back at the programmer: "Well, you're the one who taught me!" The AI has no agency here; it cannot understand unless the programmer lets it understand, and it cannot change unless the programmer gives it the tools to do so (and neither may be in the programmer's gift). Now consider the biased person. This person, when informed of their bias, is already equipped to understand, in part because they *generate* the natural language which the AI can only ever analyse mathematically. They also possess full agency to change; this person can choose to adjust how they use language or choose *not* to adjust how they use language.[21] Once a person is made aware of a bias, they may still exhibit bias, in part as a product of wider society (e.g., following social norms), and in part because of *choice*. What matters here is the latter: a person can choose to change their behaviour.[22]

To conclude this discussion of machine behaviour, I want to argue this notion could be useful for understanding what "machine-generated biases" could be. This argument hinges on the notion of choosing. In a recent article, Silver *et al.* (2021) argue that one route to achieving artificial general intelligence (AGI)[23] is to design AI systems which have sufficient means and incentive to achieve a task, while sufficient disadvantage so as to require AI systems to solve a multitude of different sub-tasks in the process.[24] This proposal raises an interesting question: what if, in the course of attempting to fulfil an objective, an AI must engage in two conflicting tasks? Such conflict may be as simple as prioritising which task to do first. In this instance, barring no prior procedure has been given by a programmer, this decision would seem to constitute something akin to a preference.[25] Alternatively, the conflict might be slightly more complicated, say, variable x takes the value of 14 when accomplishing sub-task 1, and x takes the value of 17 when accomplishing sub-task 2. What would the AGI set x to be, if x were now needed, having completed both of these sub-tasks?

I would argue that the answers to such conflicting choices *would* constitute machine-generated bias insofar as, perhaps, an AGI may prioritise the shortest task first or use whatever values reduce the computational requirements of the task (Russell, 1997). This perspective is inspired, in no small part, by human behaviour. Many economists will attest that life is a series of cost-benefit decisions, with people constantly weighing pros against cons, and preferences against preferences, to determine what they would like to do.[26] If a person must decide between eating meal A and meal B, and they cannot afford both, they must make a decision, and this decision will – in economics parlance – represent their preference. In contemporary behavioural science, this preference may also be biased, say towards whatever option appears at the top of the menu (Bucher *et al.*, 2016). If machine behaviour aims to use behavioural science to understand AI, this synthesis, which begins at the point that the AI system must make an undirected choice, may be an interesting avenue to consider.[27]

SHIFTING FOCUS

The remainder of this book will turn its gaze away from machine behaviour and towards how AI can and *is* being used in behavioural science. Yet, the ideas established here do reappear at times throughout the proceeding chapters, particularly in Chapter 5, where a key theme is the transformative interplay between AI systems and human beings. In this sense, the division I have drawn between *artificial intelligence for behavioural science*, and *behavioural science for artificial intelligence*, is probably greyer than it is black or white. But this should not change the locus of one's focus, which is largely why the division is offered.

SUMMARY: CHAPTER 1

- Machine behaviour is an emerging field which proposes using contemporary behavioural science methods to investigate the behaviour and biases and AI systems.
- The approach may adopt a perspective which sees AI systems as comparable entities to humans, though the proposals may be made broader still, and the proposals do not necessarily adopt a post-humanist perspective (i.e., humans = AI).
- Machine behaviour is limited in the conclusions it can draw based on the applicability of the methods it uses and the philosophy underpinning those methods. Quite who, or what, is behaving could be a relevant practical critique. Furthermore, the choice to use behavioural science approaches, rather than alternative approaches such as psychoanalysis, may be a point of contention (and *vice versa*).

NOTES

1. "In many settings, the only factors that are publicly observable about industrial AI systems are their inputs and outputs" (Rahwan *et al.*, 2019, p. 478). One interesting example of this can be seen on the video sharing site YouTube, where creators frequently discuss how to get videos recommended to other users (i.e., "go viral"). Two examples, which have 3.9 and 1.8 million views, respectively, are, "My Video Went Viral. Here's Why" by the creator Veritasium (2019), and "Why The YouTube Algorithm Will Always Be a Mystery" by Tom Scott (2017). Neither of these creators provide details of the YouTube algorithm, but instead provide discussions of the collective knowledge of the algorithm discerned by tens of thousands of creators uploading videos (inputs) and observing virality (outputs).
2. Possati (2020, p. 3) identifies several combinations of hybrid system: "(a) the behavior of a single AI system, (b) the behavior of several AI systems that interact (without considering humans), (c) the interaction between AI systems and humans," which can be further broken down into, "c.1) how AI systems influence human behavior, c.2) how humans influence AI systems behavior, c.3) how humans and AI systems are connected within complex hybrid systems."

3. There are several examples which could be drawn upon to illustrate point (4). For instance, on the unpredictability of hybrid systems interacting with humans, consider the incident of Tay.AI, an AI chatbot created by Microsoft in 2016 and hosted via a Twitter account (@TayandYou). The bot was shut down within a day of launching as users discovered they could manipulate the learning processes of the bot, causing it to tweet several concerning statements, including racist comments (Vincent, 2016). Note that users did not *understand* how Tay.AI worked. In many ways, they were demonstrating an inferential approach to which I will soon turn. But they understood that the bot was using tweets directed at it to learn how to converse, and this was sufficient for Twitter users to kibosh the bot.

4. Machine behaviour also seems closely related to the explainable AI (XAI) movement, and much of what is said here could contribute to that discussion too. Yet, this is a growing conversation. As Adadi and Berrada (2018, p. 52142) note in their review of the XAI literature, "Based on the conducted analysis, ideas from social science and human behavior are not sufficiently visible in this field." Also see Langer *et al.* (2021).

5. To an extent, this extends the difficulties in understanding AI systems, as the inference process focuses on *reducing* possible rules rather than necessarily *identifying* valid rules. By way of an example, the reader is invited to consider whether they would prefer to know a million rules which are not valid, or a single rule which is always valid. See van Fraassen (1989) and Lipton (2004). Furthermore, in many instances, there is some uncertainty in systems, e.g., $f(x) = 2x+1+\varepsilon$; $f(x) = x+3+\varepsilon$; where ε is some *error* or *unknown* quantity. In this case, our predictions are *probabilistic*, based on however one estimates ε .

6. Haig (2014) explicitly argues that the method of behavioural science is abductive.

7. By this I mean to say that contemporary behavioural science entertains a multitude of theories which cannot necessarily be falsified to explain observed behavioural phenomena, but still follows falsification from, say, a statistical perspective insofar as it builds on hypothesis testing.

8. I base this assessment both from a reviewer comment to this effect, and from Páez's (2019) description of the pragmatic turn in XAI. On the latter, the pragmatic perspective is that it does not matter who an understanding pertains to, so long as the understanding is sufficient to the goal in mind. Thus, one might say it matters little whether a machine is *actually* behaving, or whether such "behaviour" is an artefact of human behaviour (e.g., a programmer's choice of code), so long as the description is a sufficient description of what *seems* to be the AI system's behaviour.

9. Insofar as processing time impacts the rate at which these machines receive feedback (i.e., learn), and insofar as both machines "behave" within the same environment, such a difference may – with some parallel to genetics and experience – lead to "behavioural" differences in machines. This is an interesting idea, though beyond the present discussion.

10. Put simply, words become vectorised based on their associations with other words. This association is modelled using an *n*-dimensional space, where *n* is the length of the vector, and individual values within the vector represent coordinates. For instance, the word APPLE may have a vector of [0,1], the word ORANGE may have a vector of [0,–1], and the word FRUIT may have a vector of [0,0]. If these coordinates were to be plotted on a graph, APPLE and ORANGE would

appear further apart than either APPLE or ORANGE would be from FRUIT. This makes intuitive sense; apples and oranges are different fruits, and so have different associations and uses in natural language. But both *are fruits*, and so each has a closer association with the word FRUIT. If one were to add another word, say IPHONE, we might guess this word would have a vector such as [0,2] or [1,1], indicating it has a closer association to the word APPLE than to the words FRUIT or ORANGE. Such a vector-space formalisation is known as *semantic spacing*.

11. See previous footnote.

12. This has led some scholars (Dourish, 2016; Seaver, 2017) to begin studying *algorithmic culture*, or the sociology of communities which imagine and design AI, as a means of understanding AI technologies, rather than AI itself. I am grateful to Luca Possati for this comment.

13. This is not an argument against this approach to the study of AI systems, merely an argument that – in the process of investigation – *what* is investigated, and *how* the results are understood become products of the language, perspective, and understanding of those engaged in the investigation (Kuhn, 2012 [1962]). For example, a common practice dataset in neural network programming is the MNIST dataset, a collection of labelled images of hand-written numbers. It is not uncommon for guides which use MNIST to talk of the network learning what a loop is, or a straight line is, or learning that two loops make an 8 but a loop and a straight line make a 9. Of course, the network *does not actually learn what a loop is*, but for humans trying to rationalise the computational process, such language has some appeal. See Watson (2019) and Salles, Evers, and Farisco (2020). Also see Hofstadter (2000 [1979], p. 51, original emphasis): "[S]ymbols *of a formal system, though initially without meaning, cannot avoiding taking on 'meaning' of sorts, at least if an isomorphism is found.*" By *isomorphism*, Hofstadter (2000 [1979]) means a comparison, in this case, between machines and humans. A reader should also consider Minsky and Papert's (2017 [1969]) discussion of the difficulty of logical descriptions.

14. Possati (2020) offers a four-point schema for how humans come to determine the "set of circuits and logic" as AI. Broadly, (1) humans must set out to build AI; (2) humans must draw conclusions about how the machine must function to constitute AI; (3) humans must identify functional qualities which correspond to their ideas of intelligence; and (4) humans must be able to see themselves within the functioning system (after all, if humans are intelligent, must not the AI have some human resemblance?). While each point is interesting insofar as Possati (2020) places human interpretation at the heart of the discussion, point (3) is particularly interesting, insofar as Possati (2020) argues that something constitutes AI only if *humans think it is an AI*. If a programme does something, but this thing does not correspond to what humans *think* is intelligent, "humans go back to the project and re-start" (p. 10).

15. For instance, consider the shift from rule-based AI to node-linked AI during the 1980s–1990s (Russell, 1997; Simon, 1981; Turkle, 1988).

16. Which is not, necessarily, what the programmer *thinks* the machine has been told.

17. See Chapter 2 for a summary of connectionism.

18. In another section, which relates somewhat to the above discussion of life-hacking: "The influence of the computer on how hackers and hobbyists saw their

own psychologies was personal, and it stayed with the individual. But when the AI scientists talks about program [*sic*], it is no longer as personal metaphor. Artificial intelligence has invaded the field of psychology. As it has done so, it has built theories in which the idea of mind as program occupies center stage. And these theories have begun to move out beyond this computer culture to influence wider circles" (Turkle, 2004 [1984], p. 222). Also see Simon (1977, p. 1191) "Perhaps the greatest significance of the computer lies in its impact on Man's view of himself. No longer accepting the geocentric view of the universe, he now begins to learn that mind, too, is a phenomenon of nature, explainable in terms of simple mechanisms. Thus the computer aids him to obey, for the first time, the ancient injunction, 'Know thyself.'"

19. (Possati, 2020, p. 10): "AI is an interpreted and interpretive technology; this is the "mirror effect." Furthermore, the mirror effect is also subject to a human interpretation. This triggers a new cycle of interpretation and identification. For instance, humans can think of themselves as machines. Current cognitive science that encompasses AI, psychology, neuroscience, linguistics, philosophy and other related disciplines think of the human mind as a computing machine. This means that humans have also delegated some of the qualities and capabilities of machines to themselves."

20. Possati (personal correspondence) notes that his perspective is different to Turkle's in both the method and the outlook, with Turkle – in my opinion – being much more human-centric compared to Possati. I include this note to acknowledge Possati's position, in the absence of an adequate number of pages to express the differences appropriately. See Possati (2021) for more.

21. See Watson (2019) for an interesting discussion of the processing differences between humans and even the most advanced of AI systems.

22. I draw this perspective, in part, from de Vos (2020), who argues a key distinction between humans and machines is the ability of the former to imagine themselves as being different, something the cyberneticist Beer (1979 [1966], p. 100) has also made illusions to: "Let us begin by highlighting a particular aspect of man's reasoning faculty. It is the ability to make a forecast, and indeed it is precisely this capability in man which seems most clearly to distinguish him from other animals." These ideas are not without importance; the ability for AI to reason hypothetically remains an outstanding challenge within the field.

23. There are two types of AI: general and narrow. Artificial general intelligence would be an AI which can accomplish multiple, different tasks, while narrow AI is capable of accomplishing only specific tasks. Currently, only narrow AI exists.

24. Silver *et al.* (2021, p. 8): "General intelligence, of the sort possessed by humans and perhaps also other animals, may be defined as the ability to achieve a variety of goals in different contexts."

25. Recall that I am assuming an AGI here, and thus I also assume the system is intelligent enough to not simply return an error, but to try and find a solution to the conflict.

26. Many more economists, even if they do not admit it, will operate as if this is the case.

27. Turkle (1988, p. 259): "To function coherently, according to Minsky, an intelligent system must develop a certain inattention to its contradictory agent voices."

{Definitions: "..."} 2

INTRODUCTION

It is difficult to discuss concepts without having first unpacked what those concepts mean. To an extent, this was a difficulty prevalent in the previous chapter. For instance, in Chapter 1, I posed the question "can machines behave?" in a variety of ways, and generally conclude the answer is, "no." Yet, in this chapter, after having defined what is meant by *behaviour* and *behaving*, I would contend the answer instead is, "yes."

Yet, this is not a conclusive response. The definitions I present in this chapter are broad, and purposely so, for two reasons. *Firstly*, if these definitions are designed to help the reader understand what artificial intelligence is, then one must produce definitions which reflect the multifaceted nature of the subject. Namely, any definitions relating to artificial intelligence must be quite broad to accommodate different AI systems.[1] *Secondly*, in defining a concept such as "behaviour" or "intelligence," it is desirable for this definition to be applicable to multiple entities, by which I generally mean humans, machines, and animals. If intelligence, for instance, is defined by the number of basic calculations performed in a minute, the AI system would surely win. Of course, most people would regard intelligence as much more than this, and thus such a definition would be unhelpful. Some of these perspectives will be discussed in this chapter.

This broadness leads to the aforementioned inconclusiveness. Even with agreed definitions, people may disagree on how to apply those definitions, for a variety of reasons (Kuhn, 2012 [1962]).[2] The most obvious instance at present is that questions of behaviour and intelligence, primarily, draw on two fields – behavioural science and computer science – which may have very different understandings of these terms, and very different cultures of understanding generally.[3] Rahwan *et al.*, (2019), for instance, describe algorithmic rules

DOI: 10.1201/9781003203315-3

within AI systems as behaviours, while behavioural scientists may see these rules, as a parallel to the human brain, much more as a cognitive processes which *result in behaviour* (e.g., whatever the algorithm outputs; Furr, 2009; von Neumann, 2000 [1958]).[4]

This is all to say that this chapter does not aim to be *definitive*, but rather *productive*: to offer definitions that, at least contained within this book, allow one to ask interesting questions, offer answers to those questions, and discuss the implications of those answers, all within a coherent framework (for lack of a better word). Without such a framework, disagreements in perspective may arise due to differences in definition or understanding, as can be seen in Chapter 1.[5]

This chapter offers definitions and discussions of three key words which will shape the remainder of this book. These words are "behaviour," "intelligence," and "machine." I will expound more on each term in due course, but I will offer some immediate rationale for their selection. As I will argue, behaviours represent possible actions, but may exist without purpose. One may engage in one action for a good reason, or a bad reason, or no reason at all.[6] Within this chapter, reasoning, broadly, constitutes intelligence. Thus, to understand *intelligent behaviour*, one cannot simply define intelligence; likewise, to understand behaviour, one cannot *assume* intelligence. As above, there is a desire to keep these definitions sufficiently broad such that they may apply to various entities.[7] Yet, this is also unhelpful insofar as one wishes to distinguish between entities, notably between humans and machines. Hence, the final definition, "machine," offers recourse to this objective.

BEHAVIOUR

One need not look far for an initial definition of behaviour. Furr (2009, p. 372) defines behaviour as:

> [V]erbal utterances or movements that are potentially available to careful observers using normal sensory processes.

This definition is tempting to adopt as it is quite straightforward. Yet, it raises two points for discussion.

Firstly, Furr (2009) requires actions (e.g., "verbal utterances or movements") to be observable – or at least *potentially* observable – by others "using normal sensory processes." By "normal sensory processes," Furr (2009) means common human senses, such as sight, hearing, and touch. For Furr

(2009), what humans cannot normally observe should not be regarded as a behaviour, hence why Furr (2009) describes behaviour as "verbal utterances or movements," specifically, rather than *actions* generally. As Furr (2009, p. 372) writes, "This [definition] excludes many internal physiological responses such as neural events and blood pressure, and it excludes external physiological responses such as blushing or sweating." This is because Furr (2009) is trying to arrive at a meaningful definition of behaviour which simultaneously limits the set of potential observations which could be called behaviour too.

However, this definition is questionable in the light of technology, including AI and other technologies (Rauthmann, 2020). For instance, an fMRI machine can observe neural events, a stopwatch can record timings down to milliseconds or more, and a mobile phone can track eye movements and subtle facial expressions (Villiappan *et al.*, 2020). As technology expands, to restrict behaviour to only that which a human can naturally observe leads to a definition not conducive with the contemporary world. As such, Furr's (2009) definition may be too restrictive, as modern technology demands a broader perspective on behaviour. Therefore, I will talk of behaviours as *actions in response to stimuli*, a perspective which is quite in keeping with much AI literature and some psychological perspectives (Hayek, 1999 [1952]; Simon, 1994 [1969], 1981; Watson, 2019).[8]

Secondly, Furr (2009) talks of actions which are, "potentially available," to be observed. A behaviour which is, "potentially available," to be observed is not the description of a behaviour which *has occurred*, or even a behaviour which *has been observed*, but rather a behaviour which *could be observed*. For instance, two people, A and B, may respond to the same stimuli with different behaviours, say one screams while the other smiles. Both screaming and smiling are behaviours when described using Furr's (2009) definition. Yet, assuming persons A and B are more or less normal, both persons *could have* smiled or screamed. In short, just because an action could have been observed, but is not observed, does not mean that the action should not be considered a behaviour. Instead, it is merely a behaviour which is not *manifested* at the time of observation. Thus, I define behaviour as *potentially observable actions in response to stimuli*.

INTELLIGENCE

I have chosen to define intelligence, only, rather than *artificial intelligence*, for two reasons. *Firstly*, I am not convinced any great benefit would come from proffering yet another definition of AI. For instance, Samoili *et al.*

(2020) recently identified some 55 definitions for artificial intelligence within the literature. While I have argued for the dangers of discussion without definition, *too much* definition also demonstrates a lack of clarity.[9] *Secondly*, the term "artificial," within the phrase "artificial intelligence," is merely a description of the subject "intelligence," and given various excellent discussions of artificiality – notably Simon's (1994 [1969]) – to linger here would distract from the subject of intelligence.[10] I do not intend to diverge from Simon's (1994 [1969]) approach to the "artificial," and indeed, my approach to intelligence as a function of learning is conducive with Simon's (1994 [1969]) account of the *artificial*.[11] It is to this approach that I will now turn my attention.

Russell (2019, p. 9) writes:[12]

> Humans are intelligent to the extent that our actions can be expected to achieve our objectives.

This definition is appealing insofar as it has clear links to the definition of behaviour given above: both definitions talk of actions, and both of uncertainties (i.e., *potentially, expected*). Yet, clearly not all actions are intelligent. Behaviours such as listening to music, waving out of a window, or hitting someone in a fight are all actions (and behaviours), but to describe these actions as intelligent requires further context. From Russell's (2019) definition, an action is understood to be intelligent if it, "can be expected to achieve our objectives." As such, listening to music for a music recital is *probably intelligent*; waving out of the window of a burning building, trying to get the attention of firefighters, is *probably intelligent*; and punching someone in a fight in order to prevent serious harm befalling oneself is *probably intelligent*.[13]

It is worthwhile reflecting, momentarily, on the probabilistic nature of the above statements. Common parlance often treats intelligent behaviour in definitive terms, for instance, "it is intelligent to save money," or, "it is intelligent to exercise." But behind these statements are various normative beliefs about *what is the right thing to do* (Dolan, 2019). Yet, consider a hypothetical person whose goal it is is to maximise how long they live. To achieve this goal, they begin exercising – something which would typically be considered an intelligent act. Yet, while exercising, they have a fatal heart attack and die. Knowing this, one might pause and reconsider whether exercising was *actually the intelligent choice*. The answer, I would assert, is still yes, but not because it is understood socially to be a *certainly good choice* (or, perhaps, a common-sense choice), but because it is a *probabilistically* successful choice, insofar as – on the balance of probabilities – exercise is more likely to have positive health outcomes which achieve the goal than negative health outcomes which do not (Ellsberg, 1961). As Newall (1982, p. 104, emphasis added) writes, "Given an agent *in the real world*, there is no guarantee at all that his knowledge will determine which actions realize which goals."[14] The role of probability in intelligent behaviour will return in later chapters. With this clarification, I see no reason to substantially diverge from Russell's (2019) definition: intelligence is defined as *selecting actions which are expected to achieve one's objectives*.

MACHINE

Most readers will agree that humans and machines are different, even if both are able to exhibit intelligent behaviour. It is important to think about some differences, because they provide a means of understanding important concepts, such as autonomy and responsibility, which will come in later chapters. As such, I now turn to the definition of a machine.

Zuboff (1988) offers a potentially useful concept in their discussion of the "smart machine." According to Zuboff (1988), a smart machine differs from a dumb machine insofar as smart machines can perform tasks, but also produce records of tasks which can subsequently be employed for other purposes, such as system optimisation and learning. Zuboff (2015, p. 76) describes this two-fold nature of smart machines as the "fundamental duality," of the smart machine.

Yet, Zuboff's (2015) fundamental duality may not be as "fundamental" as her tract on the smart machine claims. Taylorism, the concept of scientific management of production lines, long preceded smart machines such as modern computers, with Taylor and his research associates standing in factories with stop clocks and clipboards, manually recording the activities of workers (Beer, 1979 [1966]; Frischmann and Selinger, 2018). Even before the Taylorism of the early 20th century, ideas reminiscent of modern computing entities such as Big Data can be found, such as in the British Imperial administration of the 19th century, where the challenge of managing a large empire required detailed records of millions of processes (Richards, 1993).[15]

Cast in a historical context, the smart machine does not appear as compelling as it does in isolation, with the meaningful difference between the smart machine and the record-makers of previous eras simply being that the human component of record-making is removed by the introduction of machinery. Nevertheless, progress has been made. If one is seeking to understand the difference between a human and a machine, one may now ask two more questions to reach an answer. *Firstly, how* is it possible to replace a human with a machine? *Secondly, why* would anyone want to?

Marx (2013 [1867]) provides answers to both questions.[16] Unrelated to the question of data which motivates Zuboff's (1988) perspective, Marx (2013 [1867], p. 257) draws a clear distinction between a tool and a machine, arguing that a tool is an entity of production which must be given its, "motive power," by a human. For instance, a human picking up and using a hammer; without the human, the hammer does not hammer a nail. By contrast, a machine possesses its own motive power. For instance, a hydraulic press as part of a sophisticated assembly line will flatten the nail irrespective of human involvement.[17] The notion of motive power is conducive with Zuboff's (1988) smart machine also; to return to the above discussion, in the days of the British Empire or Taylor's scientific management, administrators and researchers used tools such as pencils, timers, and clipboards to record data, and provided these tools with motive power. So-called smart machines such as computers displaced (automated) humans because these machines, *by definition*, possessed their own motive power to make records.[18] This, then, explains *how* it is possible to replace a human with a machine.[19] I will return to the question of motive power in Chapter 3.

While the question of motive power is yet another equivalency between humans and machines during a discussion which is meant to draw distinction, the conditions which govern this motive power differ between humans and machines. A key example which comes to mind is the need for humans to eat and to sleep. This is to say, humans are *organic entities*. For Marx (2013 [1867]), no machines would emerge unless the motive power possessed by those machines could be employed more efficiently than that possessed by humans, and thus machines are *materially different*: machines do not eat, and they do not need to sleep (Duranty and Corbin, 2022; Spencer, 2018).[20] Thus, machines are *inorganic entities*. In Chapter 3, I will expand on the notion of using AI to engage in computational tasks which are beyond the natural abilities of humans, but one illustrative example may be immediately worthwhile.

It is common within behavioural science to describe humans as *boundedly rational* insofar as humans make maximising decisions within constrained computational circumstances. Simon (1955) originally offered this term but was more interested in the notion of "chunks" (Simon, 1981, p. 366) than the idea of bias which would ultimately define contemporary behavioural science (Heukelom, 2012; Tversky and Kahneman, 1974). The theory of chunks suggests that humans can only hold a limited number of pieces of information (i.e., chunks) in short-term memory at any time (Miller, 1956; Newall and Simon, 1972) – Simon (1981) offers an estimate of seven, as does Miller (1956). However, chunks stored in long-term memory may be used to supplement chunk-calculations.[21] What constitutes a chunk is somewhat undefined, and likely differs from person to person (as does the number of chunks).[22] For the purposes of this discussion, an exact definition of a chunk does not matter.

What matters is the idea that humans are limited to only around seven chunks in short-term memory.

Now consider the computing concept of *recursion*. Recursion is a kind of self-reference. For instance, it is possible to write a computational function *A* which uses the same function *A* (i.e., itself) within it. This is not necessarily intuitive but is widely used in computing (Hofstadter, 2000 [1979]). Recursive functions can easily break because self-reference is an infinite process, and computers quickly run out of computational memory to process the never-ending function – so-called *stack overflow*. The solution to this problem is to give the computer a starting point. For instance, if a function *A* does function *A* to input *x*, function *A* might also contain a condition that says when $x = 1$, do not do *A*, but instead do *B*, where *B* is non-recursive. This means that, for any value of *x*, the computer will work backwards, pausing the previous calculation until it arrives at the value $x = 1$, at which point the initial function *A* will be defined without an *undefined* reference to itself, and all other calculations can then be performed.[23] In the meantime, these paused calculations are put into a *stack*, perhaps imagined as a tower, waiting in the system's memory to be processed. If these are not processed, the memory is exhausted, and the stack *overflows*, producing an error.

Stacks and chunks are conceptually very similar. In somewhat loose language, both essentially describe stored pieces of information which are waiting in memory to be used for something and can later be forgotten. But while humans may only possess around half a dozen chunks, and – at the time of writing – cannot "upgrade" their "hardware" to accommodate more chunks, basic computers can hold thousands of pieces of information within a stack and can easily be upgraded to accommodate larger stacks. In this simple example, a commonality between humans and machines exists, but the difference between the organic and the inorganic, or the *natural* and the *artificial*, is materially significant.[24] These definitions, and this distinction, are shown in Figure 2.1.

SUMMARY: CHAPTER 2

- Behaviour corresponds to actions which are available given what can be observed (e.g., stimuli, inputs).
- Intelligence corresponds to how one selects actions. An intelligent entity should select actions in conjunction with their objectives. Those actions which correspond to one's objectives may not be certain, however, and so there is a probabilistic component to

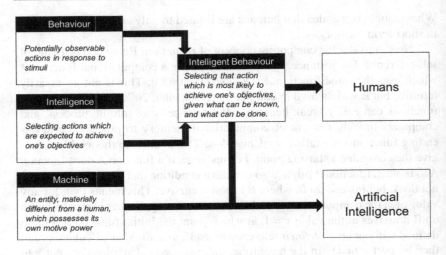

FIGURE 2.1   Definitions, and how they relate

intelligence. Intelligent behaviour can be understood as choosing actions *most likely* to achieve one's objectives, given what can be observed, and what can be done.

- Artificial intelligence is a kind of machine because (a) it possesses its own "motive power," and (b) because it is materially different to a human. Humans can be understood as an organic kind of motive power, whereas machines (and so AI) can be understood as an inorganic kind of motive power.

NOTES

1. As well as changing ideas about what AI is, and its constitutive parts, as seen in the arguments of Possati (2020) and Turkle (2004 [1984]) in the previous chapter.
2. Kuhn (2012 [1962], p. 113): "What a man sees depends both upon what he looks at and also upon what his previous visual-conceptual experience has taught him to see."
3. Kuhn (2012 [1962], p. 109): "[When] two scientific schools disagree about what is a problem and what a solution, they will inevitably talk through each other when debating the relative merits of their respective paradigms." On the specific challenge of defining concepts, Podsakoff *et al.* (2016) argue the malleability of the social sciences leads to a lack of clear division, with multiple fields claiming

intellectual ownership of multiple ideas, and multiple fields studying the same idea independently, and thus originating unique (and perhaps incompatible) lexicons and perspectives about the same thing.

4. For an example within behavioural science: Heukelom (2012) argues, from a historical perspective, that the term "heuristic" (i.e., a mental shortcut) was used very differently by Herbert Simon in the 1950–1960s, compared to Kahneman and Tversky in the 1970s onwards. For Simon, a heuristic was a set of rules which *produced* a decision. For Kahneman and Tversky, a heuristic is a set of simplifying rules which *make it easier* to make a decision.

5. A more logical mind may question why I have chosen to order these chapters in such a way. Presumably, one might argue, had the preceding chapter actually *followed* this chapter, such disagreement could have been avoided? Yet, the structure I have adopted is deliberate. I desire to present ideas *as they actually are*, rather than as a series of consensus-statements. Conflicts and disagreements arise, within and between fields, and it is perhaps just as interesting and informative to see such conflicts and disagreements as it is to digest the ideas upon which they are based.

6. This chapter will leave aside normative discussion of actions (i.e., good or bad) and focus much more on probabilistic interpretations, though later chapters will consider some normative aspects.

7. Throughout this chapter, I will use the word "entities" to describe humans, machines, and animals. Where I am talking about a *specific* entity, I will use a specific name (e.g., machine).

8. The perspective in question is that of *connectionism* based around a stimulus-impulse-response (SIR) model of the brain. By "stimuli," I generally mean any external event which causes a physical response within a person, which could be considered the internal environment. See Hayek (1999 [1952]), Minsky and Papert (2017 [1969]), and Simon (1994 [1969]). This could raise debate surrounding conscious versus unconscious responses which may also have implications for the discussion of consciousness within AI (Russell, 2019). This chapter and the rest of this book will occasionally touch on ideas which could contribute to this discussion, but it will not generally feature as a significant discussion here. Nevertheless, insofar as this discussion concerns behaviour, it may be worthwhile passing comment on the classification of some unconscious, or *myopic*, behaviours, such as breathing or the beating of the human heart. These cases are potentially oversights of the definition of behaviour I offer, though I also defer to the ideas of Simon (1994 [1969]) insofar as I consider such myopic actions necessary for the survival to sit within the realm of the *natural sciences*, whereas those actions which are non-myopic can sit within the realm of the *artificial sciences*. This is to say, the former should be described in terms of biology and chemistry and so on; the latter could be described in terms of social science, such as *behavioural science*.

9. Russell (1997, p. 57): "AI is a field whose ultimate goal has often been somewhat ill-defined and subject to dispute. Some researchers aim to emulate human cognition, others aim at the creation of intelligence without concern for human characteristics, and still others aim to create useful artifacts without concern for abstract notions of intelligence."

10. Simon (1994 [1969], pp. 4–6): "The world we live in today is much more a man-made, or artificial, world than it is a natural world. Almost every element in our environment shows evidence of man's artifice … [but] we must be careful about equating 'biological' with 'natural.' A forest may be a phenomenon of nature; a farm certainly is not. The very species upon which man depends for his food – his corn and his cattle – are artifacts of his ingenuity … [Y]ou will have to understand me as using 'artificial' in as neutral a sense as possible, as meaning man-made as opposed to natural."

11. Simon (1994 [1969]) grounds his perspective in the notion of a system of symbols. Such examples given by Simon (1994 [1969]) are spoken and written language, mathematics, and computer languages. One might add to this list musical notation (Hofstadter, 2000 [1979]). Also see von Neumann (2000 [1958], p. 81). Simon's (1994 [1969]) argument within *The Sciences of the Artificial* is that various tasks can be understood without returning to natural science principles, such as particles and molecules, but instead relying only on representative symbols (a system of symbols). This constitutes an *artificial* science. For instance, one may understand the processes of a computer with good accuracy by understanding the programming language of the computer and does not necessarily need to understand the science of the electrons which flow through the computer's circuitry. Another example: one can understand the composition of a piece of music without understanding sound wave mechanics.

12. I have generally sort to streamline the narrative here by drawing on particular, direct, quotations rather than labouring on several definitions which convey similar ideas. For the interested reader, see McCarthy (2007) and Silver *et al.* (2021) for discussions of intelligence from an AI perspective, and Sternberg (1999) for a discussion from a broader, psychological perspective. The main conflict between these perspectives may be understood by what Sternberg (1999, p. 292) laments as, "the notion of a general factor of intelligence," namely, the idea that intelligence can boiled down to a single entity (say, IQ). Yet, this conflict is not as clear cut as a plural view of intelligence versus a general view. As Silver *et al.* (2021, e. 103535) write: "What could drive agents (natural or artificial) to behave intelligently in such a diverse variety of ways? One possible answer is that each arises from the pursuit of a goal that is designed specifically to elicit that ability." Also see McCarthy (2007, p. 2): "Intelligence is the computational part of the ability to achieve goals in the world. Varying kinds and degrees of intelligence occur in people, many animals and some machines." Here, there is clearly a recognition that *what is intelligent* can be multifaceted. This, in part, contributes to what Possati (2020) and Turkle (2004 [1984]) have recognised as a transiency problem: what is considered intelligent changes depending on time and perspective. Yet, from a computational perspective, intelligence must necessarily be understood as a singular idea, in the case of Silver *et al.* (2021), *whatever actions achieve a goal*. Indeed, even Sternberg (1999, p. 292) ultimately settles on advocating for a programme of "successful intelligence," insofar as there is a recognition that intelligence is goal orientated.

13. It is interesting to note that the scenarios described, based on the perspective Russell (2019) offers, more or less represent intelligence as a *rational* action, by which one may take the broadest definition of rational as *doing something with reason*, or *doing something based on objective principles* (e.g., if I want *x*,

I must do *y*; even if I do not *want x*, if I did, then *y* would be the rational thing to do). Indeed, Russell (1997, p. 59, original emphasis) writes, "Rational agents, loosely speaking, are agents whose actions make sense from the point of view of the information possessed by the agent and its goals." See Schafer (2018) for a comprehensive history of reason and rationality. It may not necessarily be intelligent to initiate a fight and hit someone; but if the objective is simply to hurt that person, then hitting that person is an intelligent action insofar as it is done to achieve the objective. In this sense, it is also rational. Of course, this is a rather abstract view of the world, necessarily ignoring other aspects of the action, such as legal consequences, hence why subsequent iterations of rationality have often moved away from this basic understanding (Schafer, 2018). One may object, however, to the idea that intelligent = rational, but such objection is often based in one's conception of *rationality*. As Russell (1997, p. 58) notes, "The difficulty with the 'creation of intelligence' view [the idea that the goal of AI is to create intelligence, whatever this is], however, is that it presupposes that we have some productive notion of what intelligence is." Russell (1997) thus argues that rationality serves as the basis for what is understood to be intelligent. Russell (1997) further notes, as this footnote began by noting, that there are many different views of *rationality*, leading to a plurality of views of intelligence. If rationality is defined using the standard rational economic agent, *homo economicus*, then – argues Russell (1997) – even computers are irrational, and nothing is intelligent, because nothing has the unlimited processing power of *homo economicus*. Such perspectives inspired Simon's (1955) notion of *bounded rationality*, which describes maximising outcomes within constrained parameters (e.g., limited computing power). If rationality is understood as being boundedly rational, computers, humans, and animals can all exhibit intelligence. See McCarthy (2007).

14. One could define this perspective formally. Let *X* be a set of behaviours {*a, b, c ... n*} and *Y* be a set of probability functions which map perfectly onto set *X*, i.e., {$P_a, P_b, P_c, ... P_n$} The probability function is the chance of achieving some goal *Z*. One could thus define intelligent behaviour as the exhibition of whatever behaviour in set *X* corresponds to the greatest probability value in set *Y*. Yet, beyond being rather abstract, such formal definition is unhelpful insofar as it assumes sets such as *X* and *Y* actually exist, or at the very least *can be known*. For instance, the future is always uncertain, and any probability estimate of a future event necessarily misses some uncertainty. Such a formalisation, therefore, is not an accurate reflection of reality. This can also be seen by a basic inquiry into how humans *actually behave*. Few of us act based on any *conscious* estimate of the future, accurate or not. Humans are more instinctive than this, and there is good reason to be (Kahneman, 2011). As Ellsberg (1961) argues in a discussion of risk, uncertainty, and ambiguity, in many cases we do not *need to know* the probability. For instance, if one were placing a bet, one would not need to know if outcome A were 51% likely to occur or 99% likely to occur, to know the intelligent bet (i.e., behaviour) would be to bet on outcome A if the odds are always over 50%. Likewise, we do not know the probability that the sun will rise in the morning, but we know it is so likely to occur that we can proceed *as if* it is certain. Such probabilistic formalisation does not capture this nuance of reality. Yet, to talk in terms of certainties, as I have argued above, may also lead one to error.

15. It may be argued that these examples do not reflect Zuboff's (1988, 2015) smart machine, as, say, the Taylorists were engaged only in the act of record-making, and not simultaneously *doing* (whatever that may be) and record-making. However, a recent argument by Sætra and Mills (2021) asserts that, in most enterprises, the role of a manager or a foreman consists in two parts. The first is to monitor the workers (the *doing*), while through monitoring, the second role – that of the Taylorist – is thus engaged. Sætra and Mills (2021) call these two roles, respectively, surveillance as a mode of control and surveillance as a mode of production.

16. Also see Gunkel (2017).

17. Marx (2013 [1867], pp. 256–257): "Mathematicians and mechanicians, and in this they are followed by a few English economists, call a tool a simple machine, and a machine a complex tool. They see no essential difference between them, and can even give the name of the machine to the simple mechanical powers, the lever, the inclined plane, the screw, the wedge, etc. As a matter of fact, every machine is a combination of those simple powers, no matter how they may be disguised. From the economic standpoint this explanation is worth nothing, because the historical element is wanting. Another explanation of the difference between tool and machine is that in the case of a tool, man is the motive power, while the motive power of a machine is something different from man, as, for instance, an animal, water, wind, and so on."

18. Indeed, the term "computer" was – until relatively recently – a *human* job title (Becker, 2017).

19. It is curious to note that Marx (2013 [1867]) did not see the motive power of animals as anything of particular interest, and indeed, saw no problem with the motive power of a machine coming from an animal as much as it may have come from the flow of a river or a gust of wind. It is also curious to note that, for some, this whole discussion is somewhat silly. For instance, if a machine must be *switched on*, is this not humans giving the machine motive power? Turkle (2004 [1984]) offers recourse to these comments. Turkle (2004 [1984]) also follows Marx's (2013 [1867]) distinction between a tool and a machine, but notes that once machines enter into our lives, they influence our lives as we influence "their's." Turkle (2004 [1984], p. 159): "Tools are extensions of their users; machines impose their own rhythm, their rules, on the people who work with them, to the point where it is no longer clear who or what is being used. We work to the rhythm of machines – physical machines or the bureaucratic machinery of corporate structures, the 'system.' We work at rhythms that we do not experience as our own" (also see Heidegger, 2010 [1953]). Perhaps some readers will have sympathy with this perspective; some will not. But the question of "switching the machine off," is an interesting one. Consider two more cases, one computational, another social. The computational is sometimes called the "big red button problem." It follows from the question that, if one designs a sufficiently intelligent AI, how can humans ever switch the AI off? If we assume the AI maximises some maximisation function (whatever this is), it clearly cannot do this if it is switched off. Therefore, as part of its process, it will do everything possible to prevent humans from switching it off. Of course, no such AI exists currently. So instead consider the social case: On the balance of everything, should Google be switched off? This is, of course, possible, but most people – even those with

many oppositions to Google – would say no. Google, and other internet services, are deeply embedded within the lives of billions of people. Such embedding means, even where we *can switch it off* (whatever it is), we likely will not – "we work to the rhythm of machines." Or, as Aiken (2017, p. 86) writes: "The internet is not going away. We are moving to a place where we don't have a choice but to engage … Technology has become as natural as the air we breathe, as necessary to twenty-first-century survival as the water that replenishes our bodies. It has become part of our environment."

20. Marx (2013 [1867], p. 267): "The tool, as we have seen, is not exterminated by the machine. From being a dwarf implement of the human organism, it expands and multiplies into the implement of a mechanism created by man [the machine]."

21. The example Simon (1981) offers is that of novice versus elite chess players. Novices can rarely recall more than half a dozen piece-positions when shown a chess board for a brief period of time, while Grandmasters can easily recall the entire board state. Simon (1981) argues this is because novices try and store the piece-positions in short-term chunks, and quickly run out, while Grandmasters have already memorised hundreds (if not thousands) of board states and can easily retrieve the necessary information from long-term memory.

22. Consider the question of 12 × 17. One solution is to break the question down into two easier calculations: (10 × 17) + (2 × 17). Doing this in one's head, it is obvious I must first remember these two calculations (two chunks). I can then perform the first and must remember an answer of 170. This answer replaces the first calculation. I must then perform the second and remember that answer of 34. This answer replaces the second calculation (still two chunks). I then add these answers and get my grand answer of 204 (which occupies one chunk, or, when given, may be forgotten, i.e., 0 chunks).

23. Of course, if the computer never reaches $x = 1$, maybe because that was never an option available to it, the function will eventually break.

24. Humans, of course, can engage in exercises which improve cognitive performance such as memory. But many of these techniques, such as telling a story to remember details, simply encourages humans to utilise their chunks differently, perhaps truncating information, or substituting a short-term chunk for long-term information.

The Autonomous Choice Architect 3

INTRODUCTION

I have now outlined all but one of the concepts which are needed to discuss this chapter's subject, *the autonomous choice architect*. The missing concept – *choice architecture* – is what I will turn to first in this chapter.

Choice architecture describes the environment in which choices are presented to decision-makers or the design of those presentations (Hausman and Welch, 2010; Thaler, Sunstein, and Balz, 2012). One popular example of choice architecture is the arrangement of food items within a cafeteria (Thaler, Sunstein, and Balz, 2012). Others include changing default options (Jachimowicz *et al.*, 2019; Johnson and Goldstein, 2003; Madrian and Shea, 2001)[1] and exploiting social norms (Allcott, 2011; Allcott and Rogers, 2014; Bernheim, 1994; Schultz *et al.*, 2007; Sparkman and Walton, 2017).[2] The logical masters of choice architecture, so-called *choice architects*, are said to be tasked with meaningfully architecting choices so as to influence the actions of decision-makers in an intentional way, without restricting options or significantly changing economic incentives.[3]

Choice architecture emerges as a necessary component of *nudge theory*, a subset of behavioural science which seeks to meaningfully redesign choice architecture so as to influence decision-makers based on their behavioural biases, without restricting freedom of choice. A behavioural bias is a systematic cognitive tendency. For instance, changing the default option may work because people tend to be biased towards the *status quo* (so-called status quo *bias*; Madrian and Shea, 2001), which is a tendency to stick with whatever has already been selected (Samuelson and Zeckhauser, 1988).[4] The key principle of nudging is that choice architecture interacts with these systematic biases and that changes in choice architecture can change these interactions, thus producing behavioural change (Thaler and Sunstein, 2008).

Since inception, most nudging and the architecting of choices has been performed by humans, usually in teams arranged as private consultancies or, more frequently, appendages of government (Sanders, Snijders, and

DOI: 10.1201/9781003203315-4

Hallsworth, 2018). The focus of this chapter is how the role of the choice architect is increasingly being performed by AI and other algorithmic decision-making systems.

Consider an example. Facebook's *News Feed* algorithm exhibits intelligent behaviour, as described in Chapter 2, as well as being possessive of its own motive power. In the first instance, the algorithm curates around 300 posts each day to appear in the news feed, out of an estimated 1,500 for the typical user (Luckerson, 2015), with the purpose of attracting users to the platform and maximising click-through-rate (CTR). In this sense, the *behaviour* of curation is also *intelligent*, as posts are selected which are predicted to maximise CTR. The algorithm is also capable of *learning*, with a myriad of data points used to continuously fine-tune it to respond to individuals.[5] Behavioural scientists, equipped with the concepts of nudging and choice architecture, may look at the *News Feed* algorithm slightly differently (Benartzi, 2017; Yeung, 2017). As with the cafeteria, the algorithm simply moves posts around, making them more or less salient; no post is mandated (e.g., "you must click on this") or banned (e.g., "you cannot click on this"), while the principles of curation are based in cognitive bias, namely that (1) people respond to salient (e.g., easier to access) posts than less-salient posts; and (2) people have limited cognitive resources (e.g., *bounded rationality*; Simon, 1955), and would benefit from the service if it were easier for them to find what was most likely to interest them (Luckerson, 2015). The behavioural science perspective, therefore, is that a platform like Facebook consists of choice architecture (Benartzi, 2017) and that the *News Feed* feature nudges users (Lavi, 2017).

What is curious – in addition to these points – is that no *human* directly curates the 20% of posts which are shown prominently to a user. Systems, such as *content moderation*, which parse and remove content in violation of Facebook's rules may exist, but even then, much of this is automated (only later considered by an actual human; Vincent, 2020).[6] Further still, content moderation is not choice architecture as the term is commonly understood: the post is removed, not moved, or obscured; bias plays no role in how the post is handled; neither does the possible interest of users. Insofar as Facebook and similar entities[7] *architect choices*, these entities – or rather, the AI systems they design and implement –function as *autonomous choice architects*.

To reach this conclusion, it is helpful to reconceptualise choice architecture. By reconceptualisation, I mean doing much as was done in Chapter 2, where ideas were interrogated to arrive at understandings which are applicable to both humans and machines. For instance, in this chapter I will propose a view of choice architecture as *behavioural friction*,[8] which can be described quantitatively and continuously, rather than as "default options," "social norms," "framing effects," and so on, which are clearly more qualitative, discrete, and – importantly – reflective of how human choice architects discuss choice architecture. Doing so will reveal a strange view of human behaviour. When humans describe decisions, one usually says phrases like, "Person A will choose x over y because they prefer x." This is an *individual-level* view of human behaviour. But this is not necessarily a view a machine will understand. For instance, *why*, or *on what basis*, does Person A prefer x to y? How does the machine determine this preference? A human may determine Person A will choose x because they *know* Person A from a qualitative perspective (e.g., "x is red, and red is A's favourite colour.") rather than from a quantitative perspective (e.g., "x is red, and A has chosen red items 60% of the time in the data sample."). A machine will often require a quantitative perspective. For instance, if the machine knows that 60 times out of the past 100, Person A has chosen x, and 40 times out of the past 100 they have chosen y, a machine may be able to offer statements which could be interpreted as "Person A will choose x over y because they prefer x." But strictly speaking, this is not the conclusion the machine reaches. Instead, the machine's conclusion is, "Person A is 60% likely to choose x and 40% likely to choose y, based on previous choices between x and y." This is a *population-level* view of behaviour.[9]

The difference between individual-level and population-level views of behaviour is interesting. Humans rarely imagine others (or themselves) as *probabilistic* creatures, as if in the example Person A only chooses x because of some twist of fate which slightly favoured x over y. Instead, a more epistemic view is adopted – people *know* what they want and are the best positioned *to know*.[10] However, the population-level view assumes *exactly* this probabilistic character of human behaviour. Whether this view is right or wrong is much harder to determine. But it is *certainly* more interesting and should not be ignored when contending with the question of how AI fits into behavioural science.

RECONCEPTUALISATION (1): SETS OF POSSIBLE DESIGNS

I will return to this reconceptualisation later in this chapter. I instead want to begin by considering the process of *architecting choices*. As above, it is useful to arrive at an understanding of architecting choices which is applicable to both humans and machines. To put it another way, *how can the intelligent behaviour of a human choice architect be described in a way an AI system could understand?*

One place to begin is with Yeung (2017), who argues that various technologies, such as machine learning, AI, and algorithms more broadly, merely enable "selection optimisation" (p. 121) when it comes to their use within nudge-design and the architecting of choices. This is hardly an alien perspective; in the previous chapter, intelligent behaviour was characterised as the selection of (or exhibition of) a behaviour which was most likely to achieve an objective. There was also a significant emphasis on the role of *feedback*, insofar as intelligence was said to develop through a process of learning. This is how I interpret Yeung's (2017) description of *selection optimisation*, and it leads me to propose an initial description of the architecting of choices as *the task of selecting from a set of possible designs that which is predicted to be most effective at influencing the decision of a decision-maker in accordance with a pre-determined objective.*[11]

Consider a popular example of choice architecture, that of setting which option should be the default option. People tend to choose whatever option is given to them as the default option (Jachimowicz *et al.*, 2019). One striking example of this is given by Johnson and Goldstein (2003), who find that defaulting people into being organ donors (i.e., requiring them to opt-out), rather than defaulting them into *not* being organ donors (i.e., requiring them to opt-in), led to around 98–99% of people choosing to become donors, across several European countries.[12] Jachimowicz *et al.* (2019) offer many further examples in their review of the default option literature, most less dramatic than Johnson and Goldstein (2003) in effectiveness but still generally demonstrably effective. Changing the default, then, is likely a design choice which has a significant influence on decisional outcomes (Thaler and Sunstein, 2008). It is also an example which intuitively fits with the "selection optimisation" description offered above. A choice architect who is tasked with choosing which option should be set as the default option could set *any* option within the choice set $\{A,B,C \dots n\}$ as the default option.[13] Insofar as the choice architect is trying to influence people for some objective (perhaps trying to help decision-makers realise their goals, perhaps to encourage them to buy something, perhaps to produce a social benefit as is the case with organ donation),[14] the act of architecting this choice (i.e., the choice of choosing the default or choosing some other option) simply follows from *selecting* the *optimal* option to be made the default option.[15]

The example I have drawn here is for a single nudging strategy (i.e., changing the default) where the choice architect is tasked with choosing what option to nudge. But another aspect of architecting choices is determining which nudge strategy to use in the first instance (Mills, 2022; Susser, 2019). For instance, should a default option nudge be used, or a social norm nudge? Such a question may arise, say, from a difference in the propensity of bias within a given population. If the sample being targeted shows a propensity to care deeply about the opinions of others, one might hypothesise more individuals within this sample population are going to follow a social norm nudge than a default option nudge, and thus choosing to use a social norm nudge would be more effective.[16] In more general language, therefore, one can contend that the process of architecting choices by choosing the optimal choice architectural strategy is merely the process of selecting that strategy from the set of strategies $\{X,Y,Z \ldots n\}$ which is predicted to be optimal. Thus, the process of architecting choices, from both the perspective of choosing which option to nudge towards, and choosing how to nudge, can be described as a process of selection optimisation, a description which is applicable to both humans and machines (Mills and Sætra, 2022; Susser, 2019; Yeung, 2017).[17]

This is a useful description as it allows one to talk about AI systems and other algorithmic decision-making processes as choice architects. Drawing on the discussion of motive power in Chapter 2, one can then go further, and recognise that by exploiting the independent motive power of AI systems, while designing these systems to operate as choice architects, allows one to think about AI systems as *autonomous choice architects.*

A Note on Potency

Before turning to a discussion of some examples of autonomous choice architects, I want to introduce a concept which will be very important going forward (albeit in the background). In the above discussion, I have generally discussed nudging in terms of "effectiveness" or, occasionally, "optimality." In the absence of explanation of these terms, readers will likely have filled in the gaps, determining for themselves *for whom* a nudge is effective or optimal.

However, a strict nudge theory interpretation of a word such as "effective" would mean a nudge which leaves a decision-maker better off in terms of utility or welfare (Sunstein, 2014; Thaler and Sunstein, 2003, 2008). This follows from the concept of *libertarian paternalism*[18] proposed by Thaler and Sunstein (2003) and further developed in their 2008 contribution.[19] Yet, even in the absence of autonomous choice architects, techniques which "nudge" (i.e., utilise systematic bias to influence decisions without removing options) have been used for a variety of purposes, good and bad (Beggs, 2016; Lavi, 2017).[20] Indeed, even where a choice architect *thinks* they are nudging someone to be better off, because two people may be very different, one person may benefit where another person suffers (Altmann *et al.*, 2021; Mills, 2020; Sunstein, 2021b; Thunström, Gilbert, and Jones-Ritten, 2018).

It is, therefore, often unhelpful to evaluate nudging and choice architecture in *normative* terms. *Nudging for good* is often an ambition, not a strict reality. Furthermore, this is not often how choice architects *actually* evaluate nudging (Laffan, Sunstein, and Dolan, 2021). The success or failure of nudges is often discussed in terms of *potency* (i.e., how many people followed the nudge) rather than welfare (i.e., how many people benefited from the nudge). This conflict is disguised simply by assuming that nudging is good (Mills, 2020).[21] For instance, a policymaker who wants to increase retirement saving will often look at changes in retirement saving after nudging rather than whether the individuals being nudged *feel better* for having saved.[22]

An approach to this conundrum – one that helpfully reflects how choice architects *actually* evaluate nudges and how machines *are programmed* to learn – has emerged in the behavioural technology literature (Lanzing, 2019; Peer *et al.*, 2020; Yeung, 2017). This literature generally describes nudging in terms of *potency*, which should be understood to be the proportion of decision-makers who choose the option the choice architect is encouraging them to choose. For instance, recall Johnson and Goldstein (2003). In this study, changing the default option was clearly a potent nudge, as it resulted in nearly every participant becoming an organ donor. Choice architecture which leads to only, say, 10% of decision-makers choosing the architected option would not be very potent.

By thinking in terms of *potency*, and taking "effective," and "optimal," to mean potent, one can talk about the effect of changing choice architecture without necessarily having to determine the goodness or badness of the intervention. For instance, there are likely some members of society who object to nudging people to become organ donors. There are also likely many who believe this is a good policy.[23] Potency allows one to sidestep debates such as these. Potency is also useful insofar as it is easily translatable to a machine. Machines need quantitative inputs and a means for adjusting for error (i.e., learning). A nudge may very well be effective in terms of making those who follow it better off, but unless this "better-off-ness" can be quantified and used to train an AI system, this is not a useful metric. Potency is a relatively easy metric to develop an AI system and a relatively straightforward metric to programme a machine to optimise (Russell, 2019).[24] Hereinafter, the reader is encouraged to view terms such as "effective," and "optimal," as meaning potent (i.e., how many people follow the nudge).

ARCHITECTING AUTONOMOUSLY

What are some examples of autonomous choice architects? I have already offered one example – the Facebook News Feed algorithm – and this section will present several more.

The Policymaker

AI can be used to design behavioural interventions for policy. Aonghusa and Michie (2021) investigate how behavioural science can be used to achieve better public health policy. However, a vast corpus of literature concerning behavioural science interventions within public health spaces exists. This creates difficulties in two ways. *Firstly,* much of this literature draws on samples and circumstances which differ substantially. For instance, one intervention may have taken place amongst 50 elderly residents of a care home facility, while another may have taken place amongst 1,000 students at a university. The interventions being tested may have differed also, say encouraging vaccine uptake, or encouraging smokers to quit. As Aonghusa and Michie (2021) argue, these interventions may provide useful information to shape future public health policy, but spotting the important links and overlaps between studies may be difficult and non-intuitive for humans. *Secondly,* there is a lot of literature. Health is a key area of behavioural science research, and as such, many thousands of papers exist.[25] Setting aside the first issue, the time required to identify, process, and rigorously analyse each paper within the corpus would remain a significant challenge.

These issues represent an opportunity to utilise AI. Aonghusa and Michie (2021) create an AI system to extract necessary information from the corpus.[26] The AI was trained to predict the effectiveness of a behavioural intervention given various parameters which (human) users could specify. For instance, Aonghusa and Michie (2021) could tell the AI the age range of participants (say 68 to 79), the intervention type (say "Goal-Setting"), and the location of the intervention (say, a "Care Home Facility"), and the AI would output statements such as:

> For a population with **Minimum Age 68** and **Maximum Age 79**, **Goal-Setting Behaviour Change Technique** [*sic*] in a **Care Home Facility** setting is likely to lead to **5%** of subjects stopping smoking for at least 3 months.

> (p. 944, original emphasis)

A reader should see the value of such a technique quite quickly: for a policymaker with these known input parameters (say, there is a major problem of smoking amongst retirees in care homes), this system will estimate the effectiveness of a policy very quickly. This compares to commissioning a team of researchers to go through the literature and come to this (or a similar) conclusion over the course of (likely) several weeks. Furthermore, assuming the costs of each potential intervention are known, a policymaker could quickly estimate the effectiveness of *every* intervention and similarly quickly estimate the benefits of each intervention given the intervention's cost.

Of course, downsides exist. *Firstly*, the human choice architect does not necessarily *know why* the autonomous choice architect has come to the figure it has (e.g., 5% effectiveness). In many instances, some theory behind the prediction would seem desirable. *Secondly*, it is not clear what influence other studies – which a human may instinctively believe should not be used – have had on this estimate. While there are advantages to large datasets and a plurality of intervention parameters, it is difficult to determine if, say, the ten studies within the corpus set in a care home facility would be more accurate in isolation than with the addition of, say, 100 studies conducted using student populations.[27] *Thirdly*, circumstances change and may not always be reflected in studies. For instance, perhaps healthcare funding has changed, or perhaps there is a new drug which offers several healthcare benefits. The AI system does not know anything beyond what it is told; in rapidly changing or esoteric circumstances, or where the prediction draws on particularly outdated material, the prediction may be quite inaccurate, even if completely in-tune with the parameters of the AI system. *Finally*, such a system easily lends itself to being the rule, rather than the exception (e.g., "what does the AI think?"), which may be dangerous in exceptional circumstances, such as pandemics.

To an extent, however, these issues are pacified by the semi-autonomous nature of this AI choice architect. At present, in terms of implementing choice architecture, the AI is much more of a *tool* for human choice architects rather than a truly *autonomous* choice architect. There are clear instances of where the motive power of the machine displaces that of the human's, namely in the analysis of the input corpus. Yet, the means of implementing this prediction, and subsequently collecting feedback to improve the AI, remains in the hands of humans. This relationship has advantages, as intermated. For instance, the problems discussed in the immediate paragraph can be tempered by human judgement. This may actually improve outcomes– as Brynjolfsson and McAfee (2014) argue, many AI systems benefit from human–machine partnerships.[28] Therefore, it is important not to dismiss the lack of autonomy as a disadvantage.

The Retailer

Another area which is seeing extensive discussion of AI as a choice architect is online retail. Mills, Whittle, and Brown (2021) present a model of choice architecture and spending behaviour in their concept of *SpendTech*. SpendTech is defined as "a range of technologies used in conjunction with behavioural insights to induce desired spending behaviours at a decision-point" (p. 3).[29] The SpendTech concept is offered primarily as a means of conceptualising the behavioural aspects of modern e-commerce practices.[30] This is only one of several proposals which have recently emerged. Further perspectives include Villanova *et al.* (2021), who emphasise the ability of AI systems to target individuals at the optimal time to encourage purchase,[31] and Smith *et al.* (2020) who argue system's such as Amazon's recommendation AI system act as *exogeneous cognition*, thinking *for* a customer rather than encouraging the customer to think for themselves. This eases the cognitive burden for the customer (a common feature of choice architecture) and may induce easier purchasing decisions by reducing the number of decision points (Soman, Xu, and Cheema, 2010). Another important study is offered by Mele *et al.* (2021), who present novel qualitative data on how business leaders are approaching the use of these "behavioural technologies"; all of which are made viable by the computational and scaling capacity of AI.

Behavioural technologies in e-commerce, such as AI recommendation systems, use data about consumers, including past purchasing behaviour (e.g., types of products purchased, time of purchase) and likely even psychological and persuasion profiles of consumers (Kaptein and Duplinsky, 2013; Kosinski *et al.*, 2013; Kosinski, Stillwell, and Graepel, 2013; Matz *et al.*, 2017; Zarouali *et al.*, 2018), to create dynamic online retail spaces through *predicting what choice architecture will lead to sales*.[32] Further measures of individual differences (i.e., *heterogeneity*; Sunstein, 2012) amongst consumers, such as gender, age, estimated income level, and so on, could also be utilised to design an AI system to automatically deploy behaviourally informed techniques (i.e., choice architecture) to nudge consumers (Matz and Netzer, 2017; Willis, 2020). Some examples include:

- Recommendation algorithms learning from a user's previous purchases, as well as the purchases of people statistically similar to the user (i.e., collaborative filtering), to select from a set of possible products that product which is predicted to have the highest chance of being bought by the user and recommending it.[33]
- An AI system analysing purchasing habits over time to determine times within, say, a monthly cycle when a person is more inclined to splurge or more inclined to be frugal and adjust the use of choice architectural techniques such as dynamic pricing or fear-of-missing-out to maximise effectiveness.[34]
- Using (Big) data and continuous data analysis driven by feedback and reinforcement learning, AI systems identify trends and patterns between products, purchasers, and a litany of other factors (e.g., the weather) to draw inferences about customers, for instance by predicting customers who are expectant mothers, and configuring their choice environments to feature pregnancy-orientated products.[35]

Autonomous choice architects in the retail space are fully autonomous. It is not simply that the AI suggests, say, what product to put in a recommendation list before a human choice architect *actually implements the suggestion*. Instead, the AI system implements its selection itself, without any direct human oversight. This has a myriad of advantages which *are* similar to that of Aonghusa and Michie's proposal. For instance, no human *could* possibly curate a recommendations list so as to accommodate the demands of a service such as Amazon. Amazon and other large web vendors must rely on the materially different motive power of an AI.

Of course, such reliance creates risks.[36] For instance, Amazon's dynamic pricing algorithms have priced items such as books at prices as high as $2.3 million based on erroneous analysis of supply and demand signals (Parkes and Wellman, 2015), while AI systems which auto-generate product listings frequently enter into the news for the strange and amusing products they predict humans will want (Wiggers, 2020). Nevertheless, when one considers the scale at which such AI systems architect choices, one is not surprised error or strangeness occur. Furthermore, these "errors," still illustrate the advantages of an autonomous choice architect over a human in this domain. In the first instance, if dynamic pricing leads to greater profitability (as one assumes it does), then the minute and instantaneous adjustments of price represent a more efficient extraction of value than could be extracted from a human choice architect. In the second instance, if the AI system can be used to reliably identify and target specific customers – so-called *nano-targeting* (González-Cabañas *et al.*, 2021) – even products which cause confusion and hilarity for the masses may prove to be highly profitable products when targeted at a niché audience.

The Informer

An important idea, somewhat related to AI recommendation systems, is autonomous choice architects being used to curate information. Sharot and Sunstein (2020) are not directly interested in questions of technology and AI but rather a much more traditional behavioural science question: how much information should be shown to a person to assist in their decision-making? The intuitive answer would be *all information that is available*,[37] yet behavioural science ideas such as bounded rationality (Simon, 1955) suggest that such an answer would not, in fact, lead to superior decisions. Of course, the reverse – that is to say, having *no* information – would be as good as guessing, or choosing by chance, and would not lead (on average) to superior decisions either. As such, there is a persistent question within behavioural science as to how much information is best to foster deliberation without overwhelming decision-makers (Broniarczyk and Griffin, 2014; Golman and Loewenstein, 2018; Sunstein, 2020a; Thunström, 2019; Thunström *et al.*, 2016).

Sharot and Sunstein (2020) argue there are three types of utility which should be considered when determining if a piece of information should be shown to a decision-maker or obscured. These are *instrumental utility* ("Action"),[38] *hedonic utility* ("Affect"),[39] and *cognitive utility* ("Cognitive").[40] Based on these "inputs," which Sharot and Sunstein (2020) argue are subject to individual differences (in the same manner as discussed in previous examples), the subjective value of any given piece of information can be determined as a prediction: "These estimates [of utility] are integrated into a computation of the value of information" (Sharot and Sunstein, 2020, p. 16). Ultimately, this predicted value can be used to, very basically, determine if a piece of information should be shown, or, more dynamically, the *relative* value of a piece of information compared to another piece of information.

Quite what form this computation takes is not discussed by Sharot and Sunstein (2020), and it is this gap where a discussion of AI is valuable.[41] For instance, if such a calculation determines a subjective value of information, this suggests that the information presented to Person A could be different from that presented to Person B. This is information filtering, but where the redundant information to be filtered varies from person to person. Such personalisation is likely computationally intensive[42] and may lend itself more to an autonomous choice architect rather than a human choice architect (Mills, 2022). Furthermore, a choice architect (human or machine) likely has to learn *how* to determine such utility values – we rarely know *a priori* whether a person will prefer *x* over *y*, and indeed, *they may not know either.* As such, such a computational system likely benefits from constant feedback, which not only adds to the computational requirements but suggests the ability to dynamically change the information presentation would be necessary (Benartzi, 2017).[43]

It is my assertion that Sharot and Sunstein (2020) are describing, from a behavioural science perspective, various AI systems which already exist, and indeed, that one may be intimately familiar with. In the introduction to this chapter, I gave the example of the Facebook News Feed algorithm, which selects around 300 posts to show each user, out of around 1,500, each day. While one may not know *how* Facebook makes these selections, it seems reasonable to assume each post is evaluated to determine some subjective value metric for each user, and the posts which are shown are those which maximise this metric (Russell, 2019). What Sharot and Sunstein (2020) call utility, Facebook perhaps calls *engagement*. Facebook is not the only example one could draw upon; Google's search algorithm, or YouTube's recommendation algorithm, is – most basically – AI systems designed to select which information should be shown most prominently to various users.[44] Again, what Sharot and Sunstein (2020) call utility, these companies likely call *relevance*. One could also consider emerging concepts such as *robo-advice* in the financial world, where AI systems tailor financial disclosure and advice to individual users (Croxson, Feddersen, and Burke, 2019), or, more generally, what Thaler and Tucker (2013, p. 44) call a "choice engine," a hypothetical AI system utilising big data to personalise information disclosures for all manner of decisions in a person's life (also see Johnson, 2021).

While not explicitly setting out to speak to the question of AI and autonomous choice architects, Sharot and Sunstein's (2020) contribution can be seen as a behavioural science interpretation of AI systems which already architect the presentation of information for billions of people every day.

The Shapeshifter

The set of choice architecture is enormous. This, again, creates a material challenge for a human choice architect, but one which an autonomous choice architect can meet. Perhaps the clearest example of this can be seen in the field of research known as *website morphing*. Website morphing investigates the automatic, hyper-personalisation of websites (Hauser *et al.*, 2009; Hauser, Liberali, and Urban, 2014) for the purposes of influencing a user's experience on the site. For instance, Hauser *et al.* (2009) personalise websites based on cognitive styles, adjusting the information density and style depending on whether users are analytically focused, or prefer easier navigation experiences.

Yet, as Varian (2009) notes, the applications of website morphing likely go far beyond this narrow view of cognitive styles. This has also been noted by Johnson *et al.* (2012), who explicitly link website morphing to choice architecture. For instance, website morphing has been extended to morph visual aesthetics based on preferences for design elements such as colour (Bleier *et al.*, 2019; Reinecke and Gajos, 2014), and informational content based on broad cultural backgrounds (de Bellis *et al.*, 2019). In an interesting update of the website morphing concept, Hauser, Liberali, and Urban (2014) demonstrate how algorithmic analysis of a user's browsing habits, such as latent mouse hovering, click-rate, and so on, can be used to learn what interface designs a user find most engaging, and thus morph the website over time to suit these preferences.[45]

A reader may struggle to draw a clear distinction between website morphing, on the one hand, and intelligent information curation, as described by Sharot and Sunstein (2020), on the other. An example is useful here. Say two persons, A and B, are browsing a government healthcare website looking for information on the flu. Person A is young, generally fit, and does not suffer from any long-term health effects. Person B is much older, unfit relative to Person A, and has some long-term health conditions as would be expected with their age. The differences between A and B form a basis for morphing the government healthcare website to ease navigation for both persons. Yet,

Persons A and B are both suffering from the flu and therefore *require the same basic information*. For instance, both need to know the common symptoms and remedies for the flu. Yet, it makes little sense to encumber Person A with easier access to information relating to complications from long-term illness. Likewise, it may help Person B if the website is morphed so that the font size of the website is automatically larger.[46]

Website morphing is about presenting *the same information differently to different people* (Hauser *et al.*, 2009; Reinecke and Gajos, 2014). And, as with contemporary strategies to influence consumer behaviour or to curate information for more satisfying experiences, this is a task which benefits from an AI system able to autonomously change choice architecture.

RECONCEPTUALISATION (2): NUDGING, THROUGH THE EYE OF A MACHINE

I have now presented the first reconceptualisation of this chapter, namely, that the architecting of choices is the process of selecting from a set that which is most likely to achieve an objective. This allows one to imagine the role of choice architect being performed by both a human and a machine, and leads to the concept of an *autonomous choice architect*.

A key aspect of an autonomous choice architect, as demonstrated in the examples given, is that these systems are computationally more powerful, in terms of capacity to calculate and speed of calculation, than human choice architects. Thus, autonomous choice architects provide a materially different – and in this instance, superior – motive power to choice architectural design. This raises some ethical questions, some of which I will discuss later in this chapter, and some in Chapter 5.

With the second reconceptualisation, my goal is not to provide an explanation of *what* AI systems do when they take on the role of autonomous choice architects. AI systems differ from context to context and programmer to programmer, and new techniques frequently emerge which may change the optimal approach to any problem (Rahwan *et al.*, 2019). Nevertheless, computational requirements necessarily demand some transformation of concepts into models which can be understood by a machine. For instance, an AI does not *know* what a default option is; a default option to a machine is a variable which takes, for two options, one of two values,[47] say 0 or 1. An AI system also does not know what vague concepts such as *utility* or *welfare* are; it sees these things as a variable, call it *reward*, which takes a numerical value, and which it is tasked to maximise (Russell, 2019; Silver *et al.*, 2021).

I do not think it is unhelpful for the purposes of discussion to embrace the vagueness which comes from terms such as "computation," or "analysis," (or, for that matter, *black box*), nor do I think it is wholly unhelpful – in trying to explain AI systems – to anthropomorphise elements of these systems. But I do think it is helpful to *go the other way*, and to ask not how a behavioural scientist should interpret an AI, but how an AI could conceptually understand behavioural science, and choice architecture in particular. This is the second reconceptualisation I present in this chapter.

Behavioural Friction

How might a machine quantify choice architecture? The example I have given above, that of the default option, is a relatively simple example. But what of more complicated situations? For instance, say one is trying to nudge a decision-maker using a social norm nudge. This nudge takes the form of a statement with at least two varying components: the messenger and the benchmark. The messenger is who the decision-maker is being nudged to compare themselves to. This could be a neighbour, a friend, a work colleague, or just the average person in their community (Dolan *et al.*, 2012; Schultz *et al.*, 2007). The benchmark is the behaviour they are being nudged to act upon. For instance, if one is trying to reduce energy usage, the benchmark might be "used less," while the whole statement might be, "your neighbour used less energy than you." One could also add additional components, for instance, information about timespan[48] or monetary savings.[49] If, say, there are four messenger options and seven benchmark options, there are 28 combinations of this message (i.e., 4×7). Any additional variable within the statement further expands the combinations (i.e., $4 \times 7 \times n$). Should one imagine that the effect of each of these combinations will be the same? In most instances, this will not be the case (Schultz *et al.*, 2007; Sparkman and Walton, 2017).[50]

What this exercise demonstrates is that, for an AI system with many parameters which could change, it is beneficial for choice architectural designs to be classified as some sort of *continuous* quantity. Behavioural friction may be the most appropriate candidate for this quantity. Behavioural friction has become a popular topic within behavioural science in recent years as the concept of *sludge* has emerged (Thaler, 2018; Sunstein, 2019c). Broadly defined, sludge is choice architecture which slows decisions down and makes choosing options harder. For instance, excessive paperwork or difficult-to-navigate websites. Thus, it is often taken that sludge imposes friction on the decision-making process (Sunstein, 2019c). Some have distinguished between nudging and sludge from a normative perspective; nudges are good, even if they sometimes slow decisions down, where sludge is always bad (Thaler, 2018; Soman, 2020). Yet, as above, the normative perspective struggles to deal with what "good" and "bad" are, and certainly, from the perspective of a machine, this whole conversation is unhelpful.

I have argued (Mills, 2020) that one may be more mechanistic about sludge and, in turn, nudge; sludges increase behavioural friction associated with a given option, nudges decrease behavioural friction associated with a given option. In a recent contribution, Luo, Soman, and Zhao (2021) have also adopted this perspective.[51] One can see the advantages of this approach by turning to the normative perspective, where there are numerous examples of how slowing decisions down, or making choices harder to choose, can convey benefits. For instance, consider *disfluency*, a strategy whereby making text harder to read by changing the font improves understanding because it requires a reader to focus more (Diemand-Yauman, Oppenheimer, and Vaughan, 2011). Another example is food menus; it has been found that by *not* specifying which options on a menu are vegetarian, more people order vegetarian meal options (Krpan and Houtsma, 2020). This is because removing the usually helpful categories makes diners review all options on the menu, increasing the chance of them discovering an option they would have previously, automatically, dismissed. Staying in the world of food, Soman, Xu, and Cheema (2010) find that serving unhealthy snacks in smaller containers – requiring consumers to ask for refills more frequently – reduced consumption of said snacks. Finally, in the domain of saving, Soman and Zhao (2011) have found that by obscuring some savings options, people tend to save more because those options which are not obscured are thus easier to understand. These are all (arguably) examples of sludging for good; one can also think of instances where making decisions too easy (i.e., nudging) are (arguably) bad. For instance, the relatively frictionless experience of online shopping on sites such as Amazon, or the easy choices many websites offer when setting privacy options, can lead decision-makers to make choices which may not be in their best interests (Mathur *et al.*, 2019; Frischmann and Selinger, 2018). Other examples could include payday loans (Aldohni, 2021), gambling apps (Newall, 2019), and stock dealing apps (Kalda *et al.*, 2021).[52]

The relatively simple perspective – sludges increase friction, nudges decrease friction – not only avoids these debates but *is* translatable to a machine. Behavioural friction is translatable insofar as a machine can understand the *sign* of an effect (i.e., positive or negative) and the *magnitude* of an effect (i.e., its numerical value). If all choice architectural designs change the amount of behavioural friction acting on an option, a sufficiently powerful AI could understand and predict behavioural friction from information about the choice architecture (as well as the decision-maker and numerous other factors).

Inverted-U

Benartzi (2017) also adopts the perspective of nudges making decisions easier to take and subsequently argues that it is worthwhile to consider when choice architecture can make decisions harder to make.[53] This leads to, as Benartzi (2017) proposes, an inverted-U shaped curve, where some optimum comes at the point where things are neither too easy nor too difficult.[54] This, as I will show, is a very useful shape to keep in mind, though more work remains to be done. Notably, while Benartzi (2017) offers the shape and notes that considerations about the dynamic design of choice architecture are relevant when considering choice architecture within the online domain, Benartzi (2017) gives no indication of what the axes of the inverted-U shaped curve are, or how one can use this curve practically or conceptually.

Choice architecture is what choice architects *change*; as such, it is the independent variable and should sit on the *x*-axis. As above, choice architecture may be quantified for a machine in terms of behavioural friction, and therefore, the *x*-axis of this curve is a measure of friction. What of the *y*-axis? Benartzi (2017) talks in terms of optimums, which can be interpreted in several ways (i.e., optimal for *who?*). In this chapter, and in Chapter 2, I have used the perspectives of potency and probability. For the purposes of this discussion, one may see these as one and the same. For instance, a nudge may be potent if 80% of participants follow it. If another participant were to then be nudged, what would one predict their likelihood of following the nudge to be, if not 80%? Given, *firstly*, the arguments made for potency as a descriptor in this chapter; *secondly*, the arguments made for the probabilistic nature of intelligent behaviour in Chapter 2; and *thirdly*, the argument for the compatibility of these concepts now made, a good candidate for the *y*-axis value would be *the probability of choosing whatever option* is being modelled on the graph.

Figure 3.1 illustrates this inverted-U model. It shows several additional details which demonstrate the conceptual use of this model. For instance, if one begins at *F2*, the probability of choosing the modelled option (called it *n*) is given by *Xn* on the curve, which corresponds to some probability value *P1* (say, *P1* = 0.8, or an 80% chance of choosing option *n*). If one moves to *F0* by changing the choice architecture and reducing the behavioural friction associated with option *n*, the probability of choosing this option is now given by *X*, which corresponds to a greater probability value *P2* (say, *P2* = 1.0, or a 100% chance of choosing option *n*). This is *nudging*. The reverse, going from *F1* to *F0*, would be sludging, as this is increasing friction.

Figure 3.1 is simplified in several ways. *Firstly*, it only shows one option. If one assumes that within the choice set {*A,B,C … n*} is the option to choose

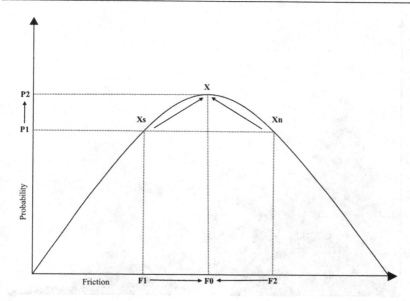

FIGURE 3.1   Inverted-U curve

nothing (e.g., in a shop, a person might deliberate between item A and item B, but the option to buy neither always remains), then for any given value of friction, the probabilities associated with each option in the set must total to 1. For instance, if the probability at point X for option n in Figure 3.1 is 100%, the probability for every option *alternative* to n *should* be 0%.[55] Figure 3.2 provides some demonstration of this on a 2d plane:

In Figure 3.2, three options are modelled. At any given value of friction, the total probabilities of all three options equal 1.0, as shown with the example values of 0.15, 0.30, and 0.55.[56]

Secondly, Figure 3.1 shows a perfectly symmetrical curve around *F0*. In reality, such a curve could be wildly skewed. The shape of the curve will reflect individual differences between decision-makers – some decision-makers will respond strongly to social norms, while others will respond strongly to shifts in time-framing or the use of colour (Mills, 2022). As such, the same change in friction (say, the changing of a website from red to green) would be expected to produce different behavioural changes in different individuals. This can be captured by changing the shape of the curve, as shown in Figure 3.3:

Figure 3.3 shows four different people, Persons A, B, C, and D, all of whom experience the same change in friction, in this instance, a nudge. However, owing to differences in the shape of the curve (because these people

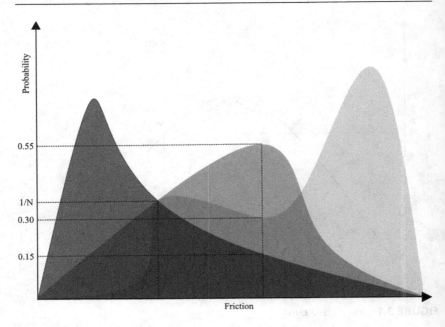

FIGURE 3.2   Inverted-U with N options (not to scale)

are different), the impact of the nudge on the probability of following the nudge is quite different. For Person A, the nudge appears to be optimal; for Person B, the nudge makes them more likely to choose the option, but the change in probability is much smaller than for Person A; for Person C, the nudge has a *negative* effect, making them *less likely* to choose this option; and for Person D, the nudge also has a negative effect, but only because the change in friction is too great, with Person D likely benefiting from a smaller change.

On the question of personalisation, which I will discuss much more in Chapter 4, these two simplifications of Figure 3.1 reveal two modes of personalisation. The first, labelled *choice personalisation* by Mills (2022), shows that choosing *which option* to nudge towards can increase potency. Option *n* may not appeal to some people, even after a change in friction, but option *m* may greatly benefit from being put front and centre. The second, labelled *delivery personalisation* by Mills (2022), shows that choosing *how to change choice architecture* may increase potency. Person C, for instance, should not have been nudged; they should have been sludged. Person A, however, definitely should have been nudged. Thus, this second reconceptualisation is congruent with the roles of the choice architect discussed above, that of selecting which options to architect, and that of selecting which strategies to use in architecting (Susser, 2019).

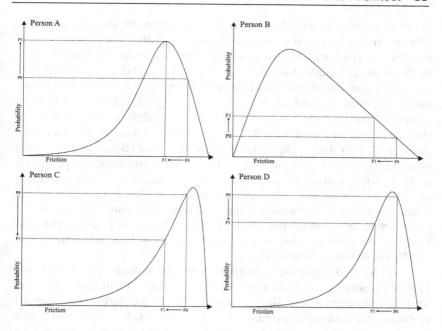

FIGURE 3.3   Different curves for different people

What's in a Shape?

Strictly speaking, the "shape" of the curve is an inaccurate description; the curve shows a *probability distribution*. As the quantity of friction, F, changes, there will be a corresponding change in the probability for the option; call it Pn. Put in mathematical terms, $Pn = f(F)$, where f is some mathematical function. As above, one cannot necessarily know what f is *a priori*; but by conceptualising choice architecture in these terms, with quantifiable inputs and outputs, one can train an AI system to estimate this function, and thus the probability distribution. Furthermore, by assigning values to each option being modelled, one can derive a maximisation function which can be used to direct the AI.

This is the second reconceptualisation in this chapter: that, for an AI, choice architecture is quantifiable friction which can be used to predict behaviour in probabilistic terms. Through this exercise, some obvious advantages can be seen. Firstly, no human could – in any timely capacity – estimate such probability distributions, especially when one considers the number of options which may be modelled (say, in the Amazon marketplace) or the number of individual differences which may be relevant (say, from an individual

Facebook profile). As with the first reconceptualisation, the material difference between the motive power of the human and the machine is a key driver of AI in influencing choices.[57] Secondly, such an approach reveals the importance of feedback and dynamism. Earlier in this chapter, I invited the reader to consider the rapid expansion of choice architectural designs as more elements became liable to be changed. Understanding so many designs is reliant on an iterative process of learning and licence to tweak designs subtly to test the effect of changes. This is a variant of A/B testing, where two designs are tried, compared, and – over time – all possible designs are ranked (Benartzi, 2017). Such processes necessarily generate a lot of data, contributing to the computational problem, but also demand choice environments which *can be changed*, often in real-time. For instance, the Facebook News Feed has an endless-scrolling capacity – if a user keeps scrolling, they will keep being shown content. Such a process requires an autonomous choice architect that is able to respond to rapid feedback (the rapid scrolling of the feed, accepting or rejecting content), and rapidly curate new content to satisfy.

But I have said much on the material difference in motive power. I am sure by now a reader will either be convinced of this thesis or will have rejected it. There are more interesting ideas to explore now. At the start of this chapter I introduced the notion of *individual-level* and *population-level* behaviour. I have now demonstrated how this difference in perspective becomes necessary for a machine to understand choice architecture and human behaviour. The shape of the curve must be estimated from data, and a single data-point will not suffice, as it might with a human. An AI requires a population of data for all manner of calculations: to determine what designs produce what quantum of friction; to reliably estimate probabilities for a single individual over time; to reliably estimate probabilities for a single individual based on statistically similar strangers; to determine statistical similarity between strangers; and so on.

Humans do not necessarily think of themselves as probabilistic nor consisting of some element of someone else. On the first point, humans may have doubts or uncertainties about what they want or should do, but once they have found a way to settle these doubts (however this is achieved), humans tend to act on the basis of *knowing*. Person A does not choose chocolate ice cream 60% of the time, and vanilla 40% of the time, because of a biased coinflip; 60% of the time, they are *certain* (i.e., 100%) they want chocolate, as they are 40% of the time with vanilla. Indeed, it is only in the hypothetical (i.e., "which ice cream will they choose next?") that this probabilistic interpretation makes some sense. On the second point, most people do not feel they are only *partly* themselves. Humans are, after all, *in-dividable individuals*. Person A may have learnt they like chocolate ice cream from Person B, who introduced them to the flavour, but that does not mean Person A's decision each time to choose chocolate ice cream is in part a function of *Person B's* decision to choose chocolate ice cream (though, of course, individuals influence one another all the time).

Nevertheless, these are the conclusions which are drawn from a population-level perspective. Basing estimates of future behaviour on past behaviour, which likely has deviances, leads one to conceptualise future behaviour as something which makes perfect sense to a machine (i.e., a probability) but may strike a human as a strange perspective. Equally, basing estimates off of the behaviours of statistically significant strangers may produce an estimate of one person's behaviour which contains artefacts of another's, which again makes perfect sense to a machine (i.e., if $x = 1$ and $y = 1$, then $x = y$), but may strike a human as strange as it draws equivalence between entities (i.e., humans) who are often held to be unique, and passionately so.

I do not know whether these perspectives on human behaviour are constructive or destructive, but they are the perspectives which come from engaging in the exercise of imagining an AI system's perspective on choice architecture and human behaviour. It reveals a point of conflict worth reflecting on. That, from one perspective, AI systems as choice architects can provide greater efficiency and personalisation and can allow humans to ask and answer much more complicated questions while responding much more rapidly to problems. But, from another perspective, such systems demand humans surrender a degree of understandability, are comfortable being subject to the efficiency and intervention of these systems, and finally, are satisfied with the vision of the human which comes to be built within the "mind" of the machine.

SUMMARY: CHAPTER 3

- The architecting of choices is a selection process, where a choice architect selects architectural components in an intelligent way. A machine capable of intelligent behaviour, such as an AI system, is therefore able to be a choice architect.
- Potency is about how many people follow the nudge. This is a useful perspective when thinking about autonomous choice architects because (a) they are given an objective, which is not necessarily to help people, and (b) AI systems require some "reward" or "optimisation" function to work.
- There are many ways AI is being used to design choice architecture, and by approaching choice architecture from the perspective of an AI system, human behaviour can be understood as probabilistic and predictable.

NOTES

1. I.e., changing the option a person would receive if they did nothing.
2. I.e., indicating to a person how other people behave.
3. Thaler, Sunstein, and Balz (2012, p. 429): "A choice architect has the responsibility for organizing the context in which people make decisions." The additional criteria I have attached in my description reflect the tenets of so-called *libertarian paternalism* (Thaler and Sunstein, 2008, 2003), which will reoccur throughout this and following chapters.
4. A suite of other biases have been proposed over several decades, including: *representativeness* (the ease at which similar situations come to mind; Kahneman and Tversky, 1972; Tversky and Kahneman, 1974); *availability* (the ease at which seemingly relevant information comes to mind; Tversky and Kahneman, 1973, 1974); *present bias* (the tendency to favour the present over the future; Laibson, 1997; O'Donoghue and Rabin, 1999a, 2015); *procrastination bias* (the tendency to delay completing tasks; Akerlof, 1991; O'Donoghue and Rabin, 1999b, 2001); *mental accounting* (the tendency to cognitively categorise resources and not deviate from these categories; Thaler, 1980, 1999, 2008; *the endowment effect* (the tendency to value that which one already has more than that which one would pay to receive it; Knetsch, 1989; Knetsch and Sinden, 1984, 1987; Kahneman, Knetsch, and Thaler, 1991); and *loss aversion* (the tendency to dislike losses more than one likes proportionate gains; Kahneman and Tversky, 1979; Tversky and Kahneman, 1992). Some of these ideas will already seem intuitive to a reader, for instance, *procrastination*, while others may appear novel. One should note that, in many instances, these biases are explanations for observed behavioural phenomena, something which leaves their existence and explanatory power to be disputed and runs the risk of encouraging all phenomena to be explained in terms of bias. For more on this perspective, see Gigerenzer (2018).
5. There will be much more discussion of personalisation in Chapter 4.
6. Aiken (2017, p. 149) has argued, owing to the sensitive and perhaps harmful nature of such moderated content, that even where humans are involved in this process, one should endeavour to give these tasks over to a machine.
7. To name a few examples: Google and its subsidiary services such as YouTube and Google Maps; Amazon; Microsoft and its subsidiary services such as LinkedIn; Facebook subsidiary services such as Instagram.
8. Understanding of this term is not immediately necessary.
9. By population, I mean based on a quantifiable sample. The example given, for instance, is based on a sample made up of observations of Person A's past choices. An alternative sample would be singular observations of the choices of 100 people similar to Person A.
10. This is something of a Kantian perspective, whereby one should assume others are reasonable and rational by default.
11. Also see Susser (2019, p. 403), whose "Adaptive Choice Architectures' are comparable to autonomous choice architecture. For instance: "[A]lgorithms

increasingly determine ... both the sets of options we choose from and the way those options are framed."

12. The counterfactual varies but is around 20–30%.

13. It seems appropriate to include a brief comment on the notion of *sets*, and set notation, as I will use sets quite frequently in this chapter.

   A set is simply a way of describing things with a common property. For instance, the set {"apple", "banana", "orange"} is a set of fruits. A set is shown using curly brackets, with everything contained within the curly brackets belonging to that set. Sets can be named formally, as in a qualitative description of what the set is (e.g., a set of fruits) or more abstractly. For instance, I could name the set of fruits N by writing N = {"apple", "banana", "orange"}. Having done this, I could then talk about this set in terms of N, or write, "set N" in order to refer to the set of fruits, and "fruits within set N" to refer to the fruits themselves. Note, the set of fruits I have given is not the set of *all* fruits. This set would be very long, but perhaps also very useful if one wanted to discuss fruits *in general*. The notation-solution used here is to add ... n to the set. n is an undefined number that should be taken to mean "all." One probably does not know all of the fruits which could go into the set of all fruits, but by writing N = {"apple", "banana", "orange", ... n} one can define N as the set of all fruits. Note that n is not a fruit; there is no fruit called n, and n does not contain whatever common property unifies all fruits. n is simply a placeholder for other fruits which would be valid inclusions in this set.

   One can see that by using sets and set notation, it is possible to discuss categories containing an enormous number of things (e.g., fruits, countries, planets, cars, choice-architectural designs) very simply, and without necessarily needing to know all the specific things which are contained within the set. (As a further note which is not of significance for this book but may weigh on the mind of a curious reader, it is possible to have *sets of sets*. For instance, if A = {1, 3, 5} and B = {1, 2, 4}, and C = {A, B}, one possible definition of C would be *the set of sets containing a* 1).

14. I will address the undefined notion of "effectiveness," shortly.

15. Of course, as will be seen, determining which option is optimal may not be simple at all.

16. Within behavioural data science, such an approach is known as persuasion profiling (Kaptein *et al.*, 2015).

17. Mills (2020) labels the selection of options *choice personalisation*, and the selection of methods *delivery personalisation*. I will elaborate more on these ideas later in this chapter, and significantly more in Chapter 4.

18. Broadly, libertarian paternalism holds people should be encouraged to pursue choices which will improve their welfare without being forced to pursue these options. This perspective is also sometimes called *soft* paternalism (Sunstein, 2013, 2014).

19. Thaler and Sunstein (2003) initially argue libertarian paternalism should leave decision-makers better off, but their 2008 contribution changes this perspective slightly, suggesting decision-makers should be better off *as judged by themselves*.

20. Beggs (2016) offers the comparison between a bank nudging a person to save, and an insurance company nudging a person to pay higher premiums. The

nudges, and broadly the industries and nature of the decision, are very similar. Yet, because saving is typically taken to be good, whereas ripping customers off is typically considered bad, these very similar situations may receive very different assessments of acceptability.

21. Note, for instance, that by the time a government has instructed a choice architect to do *something*, that *thing* has presumably been debated, negotiated, and deemed worth doing by the governmental system. The same may be true in a private organisation, though not necessarily with the same democratic legitimacy (but a legal legitimacy nevertheless). This is all to say that, in practice, it is not the role of the choice architect to debate and then do; the choice architect is the doer after *others* have debated what is to be done. In such a circumstance, it is natural to assume what is to be done is good, or at least that to do *something else* would be bad.

22. As the nudge and behavioural science programme has developed, there has been increasing work in this area. For instance, see Bourquin, Cribb, and Emmerson (2020) on the welfare effects of nudges to increase retirement saving. Also see Thunström, Gilbert, and Jones-Ritten (2018) for the welfare effects of nudges which discourage spending, and Thunström (2019) and Laffan, Sunstein, and Dolan (2021) on the welfare effects of information disclosure nudges.

23. Debates such as these are often passionate and without clear conclusions. For some perspectives, on organ donation, nudging in healthcare, and nudging in terms of manipulation more generally, see Lades and Delaney (2020), Noggle (2018), Sharif and Moorlock (2018), Simkulet (2018), Sunstein (2017) and van den Hoven (2020).

24. Indeed, if one imagines a choice architectural environment as a vector of ones and zeros, i.e., [0, 1, 1, 0, 1, 0 …], where a zero indicates a feature of the environment is not "switched on" (say, the sign-up button for a web service is *not* green) and one indicates a feature which is "switched on" (say, the sign-up button for the service is red), this vector is a wholly reasonable input for training a neural network AI, with the potency corresponding to different vectors (i.e., the percentage of people who signed up) being used as the output variable the model is trained on.

25. Aonghusa and Michie (2021) do not report the size of the corpus. This is a rough estimate by me.

26. Academic papers come in many different formats, have many different writing styles, and report many different statistics which may be useful to an analysis of this kind. To fully utilise the advantages of AI, Aonghusa and Michie (2021) had to design a programme which could automatically handle these differences in inputs. More formally, they had to specify the system's *ontology* (i.e., standard inputs).

27. These numbers are hypothetical.

28. Relatively recent discussions of the problems of estimation variance, which are broadly the problems I describe here, have been labelled by Kahneman, Sibony, and Sunstein (2021) as *noise*. They too note the potential value in improving predictions through human–machine partnerships, though from the reverse perspective, namely, that *humans* make flawed judgements and would benefit from machine guidance. Sunstein (2019a) has offered the example of judicial sentences within the criminal justice domain: judges often pass sentences for

similar crimes which vary notably, and an algorithm may be able to detect such variance and guide a judge to set more "appropriate" sentences (of course, we may not know what is appropriate; noise is an imperfect domain of study).

29. A decision-point is defined as any moment where a decision must be made (Soman, Xu, and Cheema, 2010). The SpendTech model simplifies by focusing only on a final binary decision – to buy or not to buy. Yet, if a purchasing experience can be thought of as a *series* of decision-points (e.g., the decision to go to a website, the decision to browse item 1, 2, 3, etc., the decision to add to basket, and ultimately, the decision to buy), SpendTech could be applied at *each* of these decision-points. This makes SpendTech rather broad, and more complicated (i.e., one must conceptually model many branching paths in a large decision-tree to capture the whole experience). I am grateful to my co-authors for this comment.

30. Also see Kaptein *et al.* (2018) and Kaptein and Duplinsky (2013).

31. This may be reminiscent of the Fogg model of behavioural change.

32. The perspective often adopted (e.g., Kaptein and Duplinsky, 2013; Mills, Whittle and Brown, 2021) is that such autonomous choice architects are implemented by vendors to maximise vendor interests (i.e., sales, profits). Also see Mathur *et al.* (2019) on *dark patterns*.

33. This reflects a growing tendency to see data not *individually*, but *collectively*, i.e., data exists as a combination of observations about many individuals (Viljoen, 2020). Social norms and other comparisons have long been held to be key drivers of economic behaviour (Bernheim, 1994; Veblen, 2012 [1899]). For a relatively recent study that applied such interventions to spending behaviour, see Payne *et al.* (2015).

34. The role of time has been a staple of the behavioural science toolkit for some time (pardon the pun; Service *et al.*, 2015). The basis of timely interventions remains to be further elaborated upon (for instance, a reading of the EAST document – a standard framework for applying behavioural insights – reveals the discussion of some studies that do not immediately seem tied to *time*, such as different wordings of text messages, and citations of research looking at the phenomenon of *subjective* time, rather than *timely interventions*), but is offered here following the intuition that people will be more likely to spend at or around pay day, than several weeks after pay day. Also see Villanova *et al.* (2021). For a famous study on the effects of time on decision making, see Danziger, Levav, and Avnaim-Pesso (2011) for the impact of meal timings of judicial decisions. For a behavioural framework applicable to temporal considerations.

35. See, for instance, Zuboff (2015, 2019). The example of pregnancy draws from several cases of such incidents. See Hill (2012). For some incisive criticism of this incident (namely, that the story represents selection bias, as the store in question, Target, probably sent the same coupons to many women, and secondly, that this incident tells one nothing of the accuracy of the prediction system), see Fraser (2020). For a recent discussion of *nano-targeting*, see González-Cabañas *et al.* (2021).

36. I have chosen to only discuss a risk immediately relevant to this discussion. The notion of more systemic risk proposed by having large parts of the economy reliant on AI systems is a fascinating question, and one deserving of some dedicated scholarship. For an interesting starting text, see Öhman and Aggarwal (2020).

37. This follows from the characteristic of *homo economicus* having perfect knowledge.

38. "Will the knowledge help, hinder or have no influence on my ability to make decisions to increase reward and avoid harm?"

39. "Will the information induce positive or negative feelings, or will it have no influence on my affect?"

40. "Will information improve my ability to comprehend and anticipate reality?"

41. Sharot and Sunstein (2020) *may* be aware of some computational techniques when they describe this vague computation. For instance, they choose to include a diagram of this computation which has the structure of a simple perceptron model, though this is not explicitly stated (a perceptron is an individual node within a neural network which takes various inputs, runs a computation – using a sum of weighted products bounded by some activation function – and returns a single output; Minsky and Papert, 2017 [1969]). A recent empirical examination of the theory put forth by Sharot and Sunstein (2020) does not use AI, but more traditional statistical methods. See Kelly and Sharot (2021).

42. For instance, how are the individual utility functions calculated? These calculations could, in themselves, be computationally intensive. Furthermore, adding in a suite of individual differences (e.g., age, gender, ethnicity, income-level) further increases the calculation. Finally, such calculations are to be done for each quantum of information which *could* be presented, and so the number of calculations is a product of the number of quanta of information.

43. One reading of Sharot and Sunstein's (2020) contribution would not be the one presented here. For instance, the paper could simply serve as a guide for evaluating whether a piece of information should be shown. Sunstein's (2019b) work on food calorie labels speaks less to the dynamic, digital information platforms I have alluded to and will shortly draw upon, and more to impersonal, regulatory disclosures. Equally, however, Sharot and Sunstein (2020) do *emphasise* the subjective component of this computation, which suggests one *should* read this contribution as one speaking to a dynamic, personalising information platform rather than a static, impersonal disclosure label.

44. Search prominence is also important. For instance, it has been reported some 71–92% of users do not click past the first page of Google (Shelton, 2017).

45. Website morphing is a much-discussed idea within marketing science and information systems literature studies. This is somewhat to the disadvantage of the discussion offered here, as the language of website morphing often overlaps with similar (and previously discussed) ideas such as recommendation systems and collaborative filtering. For instance, Chung *et al.* (2016) discuss the "morphing" of social media news feeds based on user reading habits and social media data. Likewise, Kosinski *et al.* (2013) and Kosinski, Stillwell, and Graepel (2013) provide evidence of how social media data can be used to predict personality traits, which in turn can be used to target advertisements (Matz *et al.*, 2017). Also see González-Cabañas *et al.* (2021) on *nano-targeting* using Facebook data, and Ferwerda and Tkalcic (2018) on predicting personality using Instagram profiles. Such studies demonstrate the potential of user data to determine *a means to morph* features of user interfaces, but do not speak strictly to the notion of website morphing.

46. Further examples abound. For instance, Person A may be happy with a text-message consultation, or receiving guidance via a smartphone app, while Person B

may prefer a telephone consultation and information on how to book an appointment with their local general medical practitioner.

47. And for n options, one of n values.

48. I.e., "your neighbour used less energy than you last month."

49. I.e., "your neighbour used less energy than you, saving them $100."

50. Indeed, one cannot know *a priori* how different combinations will relate because of spillover effects (Dolan and Galizzi, 2015). For instance, say messenger x typically *reduces* the effectiveness of the message. If the combination of messenger and benchmark is treated as additive (i.e., effect = messenger + benchmark), then messenger x would be negative. Yet, say messenger x with benchmark y is found to be extremely effective, while benchmark y in combination with other messages is pretty ordinary. An additive approach cannot explain this result. At best, one must assume deeper complexity between components. For instance, perhaps the relationship is productive (i.e., effect = messenger × benchmark), or perhaps there are feedback loops (i.e., "I don't like messenger x, but this benchmark is not something x would do unless they had a good reason, so maybe I should listen to x").

51. "A nudge is defined as an intervention to facilitate actions by minimizing friction and removing impediments, while a sludge is defined as an intervention that inhibits actions by increasing friction" (Luo, Soman, and Zhao, 2021, p. 2).

52. An astute reader might note that many of these examples appear to show elements of *nudging*. For instance, if obscuring other options makes the unobscured option easier to choose, is this not a kind of nudging? Indeed, this may be a valid interpretation – Mills (2020) argues that by making any option within a choice set easier to choose through changes in choice architecture (i.e., nudging), at least one other option within the choice set must become harder to choose (i.e., sludging), and vice versa. This is called *nudge/sludge symmetry*, and further serves to undermine the normative definitions of sludge.

53. It is from Benartzi (2017) that I draw the example of disfluency offered above. Note that Benartzi (2017) was writing prior to the emergence of sludge, and I suspect that had the term been around at the time of their writing, Benartzi (2017) would have offered a discussion in these terms.

54. This perspective is also implicit in the discussion of too much or too little information offered above.

55. There may be instances where a person can choose multiple options, and therefore this statement may not hold. There are conceptual solutions to this (e.g., make the option being modelled the option of choosing, say, *two* options). But these solutions are somewhat tedious. For the purpose of discussion, I adopt something resembling an effective demand perspective, and assume a person chooses one option, and *can only choose one option*.

56. The value of $\frac{1}{N}$ demonstrates a point of indifference between the three options. Indifference is definitionally given at the point where the probability of choosing each option within a set of N options is the same, i.e., $\frac{1}{N}$ For more on this case of indifference, see Mills and Whittle (2022).

57. Assuming, of course, such methods produce enhanced efficiencies. See Chapter 5.

AI Knows Best

4

INTRODUCTION

Throughout this book– the exception perhaps being towards the end of Chapter 3 – a key character in the story of AI and behavioural science has taken a backseat. That character is the *decision-maker*. Previous chapters have focused on the machine, on relevant definitions, and on the choice architect, but the decision-maker has been treated as a passive actor or some variable in a computational model. To an extent, this is not surprising. For instance, a key advantage of using AI in behavioural science, as discussed in Chapter 3, is a machine's ability to process more data faster than a human choice architect could. This application has little to do with the decision-maker whose choices are being architected, and indeed, this decision-maker *may never know if*, or *may never know how much*, their choice environments are the product of human or machine design.

At the same time, decision-makers cannot be excluded from this conversation. For instance, if an AI is optimised to maximise something like the click-through-rate (CTR), then the AI will engage in strategies which attempt to alter the behaviour of decision-makers. This may not simply be by identifying novel items for the decision-maker to click on which align with the decision-maker's preferences but by actually *changing* the decision-maker's preferences (Russell, 2019). Many people feel uneasy at the idea of being manipulated or having their agency constrained or disregarded (Sætra, 2019), and such debates have been mainstays of the nudge theory literature for quite some time before the emergence of the AI systems discussed in this chapter (Mitchell, 2005; Rebonato, 2014). While AI may offer computational advantages, these same advantages may exacerbate these concerns (Sætra, 2019), as well as bring new problems to the fore.[1] From these perspectives, a discussion of AI and behavioural science would not be complete without some consideration of how decision-makers interact with and respond to autonomous choice architects, as well as how AI systems change *how* decision-makers can interact and respond (Susser, 2019).

DOI: 10.1201/9781003203315-5

A central theme of this chapter will be what happens when a person tries to go against a machine's nudge. In nudge theory, a decision-maker should always be able to "go their own way" (Thaler and Sunstein, p. 5) insofar as they do not face altered economic circumstances which might serve as undue coercion.[2] Yet, a growing body of literature exploring the synthesis of various features of AI – such as reinforcement learning, big data, and dynamic feedback – with nudge theory has begun to question the ease by which AI-driven choice environments allow people to go their own way. It is this literature and the literature's central character – the *hypernudge* – which I will focus on in this chapter.

"WHAT IS A HYPERNUDGE?" PART 1

The premise of this book is to discuss how AI fits into behavioural science, and a tentative glance over the literature might encourage one to see hypernudging as the first serious attempt to engage with this discussion. Yet, the literature which deals *specifically* with this term has remained rather small, in part I believe because the term itself has struggled to be practically related to either behavioural science or computer science.[3]

Yeung (2017, p. 122) describes hypernudges as:

> nimble, unobtrusive, and highly potent, providing the data subject with a highly personalised choice environment ... Hypernudging relies on highlighting algorithmically determined correlations ... dynamically configuring the user's informational choice context in ways intentionally designed to influence decisions.[4]

This description provides some clues as to what a hypernudge is. For instance, hypernudges are algorithmic in nature, personalised to data subjects, and able to change subtly (i.e., "nimble", "unobstructive"). In many ways, the ideas which initially constitute a hypernudge should be quite familiar to the reader. Yet, this description does leave much to be inferred. For instance, the term "hypernudge" would seem to link this description to nudging, but Yeung (2017) provides almost no discussion within this description of the similarities between nudging and hypernudging. Instead, one must infer that hypernudges *are* nudges, and thus do not mandate or ban options, or "significantly change economic incentives" (Thaler and Sunstein, 2008, p. 5). Furthermore, Yeung (2017) offers relatively little discussion of the technological component in this description – personalised *how*, correlated *between who*, dynamically reconfiguring *what*? For the term hypernudge to be a useful contribution to this discussion – *as it can be* – some recourse to these questions must be offered.

Lanzing (2019) attempts to do so, offering a more regimented description of a hypernudge.[5] Following Lanzing (2019), a hypernudge consists of three elements:

- *Dynamism*, which refers to personalisation, whereby the choice architecture to which a decision-maker is exposed is personalised to that decision-maker in real-time.
- *Predictive Capacity*, which refers to systems such as algorithms predicting behaviours and learning from prediction via built-in feedback mechanisms.
- *Hiddenness*, which refers to how noticeable the extraction of data and the act of nudging are.

While Lanzing's (2019) specifications are more explicit than Yeung's (2017), and while I will generally utilise this structure within this discussion, the features Lanzing (2019) identifies are nevertheless merely clearer specifications of ideas already raised by Yeung (2017): Yeung (2017) writes of dynamic, real-time reconfiguration and personalisation throughout their work; they describe hypernudges in terms of cybernetics, receiving and responding to feedback over time; and they argue hypernudges follow from nudges insofar as both, citing Bovens (2008, p. 3), "work best in the dark." As a result, a curious reader may still be left wondering quite *what a hypernudge is*.

Some other discussions of hypernudging can be found, though these are certainly less focused on the topic than Lanzing (2019) and Yeung (2017) are. Sætra (2019) cites hypernudging in their discussion of the possibilities of big data being used to produce highly accurate predictions of human behaviour. Yet, Sætra's (2019) discussion differs from the concept of hypernudging in two crucial ways. Firstly, Sætra (2019) is concerned with big data used to make single, highly accurate predictions, avoiding discussions of inter-temporality and dynamic feedback which both Yeung (2017) and Lanzing (2019) proposed as key features of hypernudging. Secondly, Sætra (2019) is interested in how the notion of *perfectly predictable* choices impact the political theory of choice, freedom, liberty, and so on, rather than adopting a behavioural science perspective, which would be helpful here. Likewise, one can turn to Darmody and Zwick (2020), who too make reference to hypernudging – this time within the world of consumer behaviour – but more so as a broad description of *technology-driven nudging* rather than as anything distinct. Finally, a recent contribution by Smith and de Villiers-Botha (2021) discusses the use of hypernudges to influence the decisions of children, arguing that children require protection from hypernudges as their preferences and cognitive processes are still developing and should not be unduly inferred with.

This last point is important because it is where I have found most discussion ends. In the absence of specificity regarding behavioural or computational science, I have found in many discussions with colleagues hypernudging is subsequently, and *inadequately*, defined as: *hypernudge = nudging + big data*.[6] One may be satisfied with this perspective, particularly if one is convinced of the notion of an autonomous choice architect discussed in Chapter 3. But I maintain that the concept of hypernudging is something worth digging deeper into, especially if one wants to arrive at a perspective of AI and behavioural science which is orientated around the decision-maker. As such, rather than accepting this additive definition, I would urge one to stick with this question a little longer.

"WHAT IS A HYPERNUDGE?" PART 2

One cannot wholly abandon the descriptions put forth by Yeung (2017) and Lanzing (2019), and indeed, one would suffer for having done so. I will broadly follow the structure of dynamism, predictive capacity, and hiddenness given by these authors. However, I will specifically focus on dynamism for two reasons. *Firstly*, it is where I consider the key distinctions between hypernudging and nudging to be found. *Secondly*, because much of what could be said for predictive capacity has already been considered in Chapter 3, while much of what can be said of hiddenness is, in my opinion, better dealt with in an analysis of hypernudging once conceptually understood.

Dynamism

Dynamism does not just appear within discussions of hypernudges but comes up in several areas within behavioural science and has appeared in several instances in preceding chapters. For instance, *digital* nudging and online behavioural science (Benartzi, 2017; Weinmann, Schneider, and vom Brocke, 2016) regularly draw on the notion of dynamism, as argued in Chapter 3, because the enhanced motive power of an AI allows for greater processing of information and more incremental adjustment via personalisation, and because the online medium allows for more design flexibility to communicate these adjustments, compared to offline (e.g., Hauser *et al.*, 2009). These two parts represent core components of dynamism, and I want to take some time to focus on each.

Personalisation

While personalisation is a feature commonly associated with big data and technology (Benartzi, 2017; Mills, 2022; Porat and Strahilevitz, 2014; Sunstein, 2012; Yeung, 2017), the impetus for *personalised nudging* and *personalised choice architecture* comes from what Sunstein (2012, p. 6) calls, "the problem of heterogeneity." The problem of heterogeneity occurs when different (i.e., heterogeneous) individuals are nudged in the same way. Assuming the nudge has been used because it is expected to produce some welfare benefit *across the population* being nudged (i.e., a *net benefit*), because individuals within the population are different, it would be expected that some individuals greatly benefit from being nudged, while others may suffer. Thus, even when the net welfare benefit of an impersonal nudge[7] is expected to be positive (i.e., benefit > suffering), a theoretical argument could be made that the benefit of nudging could be increased if nudges could respect individual differences within the population (Mills, 2022; Peer *et al.*, 2020; Sunstein, 2012; Sunstein, 2021b).

One solution to the problem of heterogeneity offered by Sunstein (2012) is personalisation. This argument follows that if the problem of heterogeneity emerges because the same nudge is applied to a heterogeneous population, then applying many personalised nudges which respect the individual differences within the population may allow a *nudging programme* to produce greater net welfare outcomes than an impersonal nudge would.[8]

It is interesting to note that Yeung (2017) is implicitly aware of the problem of heterogeneity and the potential solution which personalisation offers. Consider this extract, where Yeung (2017, p. 122) discusses an example of a speed hump as a nudge:

> Although vehicles should proceed slowly in residential areas to ensure public safety, speed humps invariably slow down emergency vehicles responding to call-outs. In contrast, Big Data-driven nudges avoid the over- and under-inclusiveness of static forms of design-based regulation.

Here, the speed hump functions as an impersonal nudge insofar as it encourages vehicles to slow down but does not strictly coerce the driver via, say, a large fine (Thaler and Sunstein, 2008). However, not all vehicles are the same, and some – such as emergency service vehicles – would benefit from not being nudged to slow down. If it were possible to design a speed hump that did not impact emergency service vehicles, while still impacting other vehicles – in other words, to personalise the nudge – an argument could be made that net welfare has been increased: the gain of the welfare of public safety afforded by the humps remains, while the loss of welfare from slower response times, incidentally, produced by the hump, disappears.[9]

As concerns the question of personalisation, it is also interesting to note the particular relationship Yeung (2017) supposes exists between personalisation and big data, namely, that it is *big data-driven -nudges* which avoid the problem of heterogeneity through personalisation, which could imply that personalisation necessarily *requires* big data and related technologies. This is a rather common perspective. Porat and Strahilevitz (2014), Thaler and Tucker (2013) and Weinmann, Schneider, and vom Brocke (2016) all suggest this conclusion. Sunstein (2013) argues that big data and sophisticated analytical procedures will be necessary to personalise nudges, also.

There is some rationale for this equivalence between personalised nudges and technologies to personalise, and indeed, insofar as one is concerned with AI, I do not dispute the necessity of big data and significant computation.[10] Furthermore, I do not dispute the need for *some* data when personalising; in order to address the problem of heterogeneity, one must *at least* know how individuals differ and have some intuition (if not a model) for responding to this difference (Mills, 2022). Yet, as Mills (2022) argues, the *avenues* by which a nudge *can actually be personalised* are not dependent on technologies such as big data.

Mills (2022) proposes a two-component framework for personalising nudges: *delivery personalisation*, which involves selecting the *type* of nudge used (e.g., default option, social norm) based on heterogeneity data, and *choice personalisation*, which involves selecting the *option* nudged towards, based on heterogeneity data. This framework has already been seen, in two forms, in Chapter 3. In the first instance, I argued that choice architects select from two sets, one a set of nudge strategies and another a set of nudgeable options. These sets represent delivery and choice selection, respectively (albeit without the personalisation). In the second instance, I argued that the probability of choosing an option could be thought of as a curve which is navigated by changing choice architecture. In this model, the shape of the curve was determined by individual differences, while each option possessed a unique curve. Thus, the shape of the curve represents the delivery component, while the choice of which option-curve to focus upon represents the choice component.[11]

Following this framework, a nudge is personalised provided at least delivery personalisation or choice personalised is used. As such, the framework defines personalisation around the *method* of personalisation (i.e., what is changed?) rather than the *means* (i.e., how is it changed?). This leads Mills (2022, p. 150) to argue, "choice architects will require access to heterogeneous data [to personalise nudges] ... [but] such data need not take the form of big data." Porat and Strahilevitz (2014, p. 17) have dubbed such "non big data" personalised nudges as "crude" personalised nudges, which one could contrast with the "highly sophisticated" hypernudges described by Yeung (2017, p. 121).[12] See Table 4.1.

TABLE 4.1   Personalisation Definitions

TERM	DEFINITION
Choice Personalisation	The use of various heterogeneity data to determine what is the best outcome to nudge a decision-maker towards.
Delivery Personalisation	The use of various heterogeneity data to determine what is the most effective method of nudging an individual.
Crude Personalisation	The use of obvious or easily accessible heterogeneity data, in conjunction with computationally simple methods, to personalise choice architecture.
Sophisticated Personalisation	The use of abstract or difficult to access heterogeneity data (and other data), in conjunction with computationally sophisticated methods, to personalise choice architecture.

Real-time (Re)Configuration

As above, dynamism consists of two aspects, personalisation and real-time (re) configuration. Personalisation may well fulfil the promises of superior nudging in terms of welfare (or potency), but only insofar as one is *able* to personalise. As Thaler (2021) has noted, in many instances choice architects are limited by what they can practically do, either because the medium does not make itself easy to change dynamically (e.g., a paper form), or because there are various interests or pre-requisites at play (e.g., a politician telling a choice architect they *must* only change the default option). I have already briefly touched on another limitation, namely access to data. As Sunstein (2012) notes, one must be able to access data both to determine how people are different and to reliably target differences when identified. If personalisation is the theory, real-time (re)configuration is the practice.

Yeung (2017) seems aware of this. As well as describing big data driven-nudges as free from "over- and under-inclusiveness," Yeung (2017) asserts they differ from "static forms of design-based regulation," such as paper forms (Benartzi, 2017; Thaler, 2021). This seems like a substantial distinction between traditional nudges and the hypernudge concept, so much so that I assert it is *the key distinguishing feature*: hypernudges *change*.

Consider two examples discussed by Yeung (2017): the Facebook News algorithm and Google Maps GPS. On the former, depending on a myriad of factors, the content stream (i.e., the feed) displayed on the News app changes. The method of nudging itself may not change, though – following the two-component model of personalisation – it would still be consistent to say changing what content is nudged is still changing choice architecture. On the latter,

Google Maps can be used for directions when driving or navigating on-foot. As the user changes their location, and as other changes occur, such as traffic build-up, Google Maps is able to change the instructions it gives the user. Once more, the method of nudging may not change, but the outcomes being nudged towards (e.g., "turn right", "turn left") are changing, and thus, so too is the choice architecture. These changes are undertaken by the choice architect, and – as has been seen in Chapter 3 – these choice architects are AI systems which operate autonomously.

A Note on Predictive Capacity and Hiddenness

I have never been good at getting to the point, but I think I am now close to it, at least as concerns hypernudging. Before moving on to the next section and exploring the implications of the *changing* characteristic of hypernudging, I want to take a moment to address predictive capacity and hiddenness.

Much of what could be said here regarding predictive capacity has been discussed previously. Namely, choice architects (be them human or machines) choose choice architecture based on predictive effectiveness, often in terms of potency. Autonomous choice architects have a predictive advantage in many domains, and thus the process of prediction often takes on a mechanistic (or, as Yeung (2017) puts it, *cybernetic*) dynamic of objective-action-feedback, the latter of which constitutes the machine's "learning," which in turn should improve the predictions.[13]

Hiddenness is perhaps the feature which least distinguishes hypernudges from nudges. The argument follows that while nudges may work when revealed to decision-makers, "nudges work best in the dark" (Bovens, 2008, p. 3), which is to say, when a decision-maker does not realise they are being nudged. This is an interesting, and quite old, argument within nudge theory, with concerted beliefs on both sides. In recent years, a significant body of research has been produced on the matter of nudge transparency (Bang, Shu, and Weber, 2020; Bruns *et al.*, 2018; Kroese, Marchiori and de Ridder, 2016; Loewenstein *et al.*, 2015; Steffel, Williams, and Pogacar, 2016), with studies generally finding that nudges continue to be effective even when revealed to decision-makers. However, a strict interpretation of Bovens' (2008) initial statement – which directly informs the perspectives of Yeung (2017) and Lanzing (2019) – is that nudges work *best* in the dark, not that they *only* work in the dark. Thus, there may still be room for some debate. Furthermore, the literature on persuasion has often held that *who* is persuading can have a significant impact on the acceptability of the persuasion (Brown and Krishna, 2004; Friestad and Wright, 1994, 1999; Sternthal, Dholakia, and Leavitt, 1978), and it is interesting to note that the *who* in many studies of nudge transparency are those who are typically trusted in society, for instance, doctors (Loewenstein *et al.*, 2015). The apparent acceptability of nudging may fall if the agent doing the nudging was not as trusted as those used within the existing literature. Lanzing (2019) does offer a slightly different perspective on hiddenness beyond Bovens' (2008) perspective. For Lanzing (2019), a component of hiddenness which is relevant to hypernudging is the hiddenness of the motivations and incentives of choice architects. The argument follows that several properties of hypernudging, such as their *automaticity* (i.e., real-time reconfiguration) and *non-physicality*,[14] allow hypernudges to seamlessly fade into the background, robbing decision-makers of the opportunity to evaluate the acceptability of the hypernudge.[15] This is similar to what Susser (2019) describes as a philosophical understanding of hiddenness. For Susser (2019) – as for Lanzing (2019) – technologies become invisible when people stop seeing them as technologies and simply as means to an end. For instance, Facebook may become "invisible" if people use Facebook to read the news without ever reflecting on the fact *they are using Facebook*.

THREE BURDENS

If I were to offer a definition of a hypernudge, it would be thus:

> Hypernudges are systems of dynamic nudges that change over time and in response to feedback.

Hypernudges can only exist where dynamism is possible and are thus dependent on the medium with which they are used (Morozovaite, 2021) and the data which can be incorporated into the design process.[16] Having some grasp of what a hypernudge might be, I now want to turn to how hypernudges – implemented by autonomous choice architects – interact with decision-makers. Broadly, I suggest hypernudging creates three burdens for decision-makers, the first which arises when the decision-maker tries to avoid the hypernudge, the second which arises when the decision-maker tries to understand the hypernudge, and the third which arises when the decision-maker accepts the hypernudge.

The Burden of Avoidance

The burden of avoidance is a significant consequence of the dynamic nature of hypernudges. While Yeung (2017) and Lanzing (2019) implicitly touch on this burden, neither follow the burden of avoidance to its natural conclusion, especially as a contrast with traditional (i.e., static) nudging. To appreciate the burden of avoidance, it is helpful to consider two examples of nudging.

Recall the example of the speed hump as a nudge. The speed hump nudges insofar as it encourages the driver to slow down but does not force or otherwise significantly coerce the driver to slow down. The driver could choose to drive over the hump at speed, only suffering the temporary physical discomfort of doing so. In driving over the hump with speed, the driver is essentially *opting-out* of the nudge; they are choosing the option *not* nudged towards. As they have driven down that particular road containing a speed hump, the driver *must decide* to follow the nudge or not. But once this decision has been made – say, *to not* slow down – the driver is free from the hump. The hump does not chase them down the road, nor does the road itself contort to produce additional humps *in response* to the driver's decision. Another example of this, to stress the point, is opt-out organ donation. A person may be nudged, say when passing their driving test, to become an organ donor, with the default option set to "Yes, I want to be an organ donor." This is quite a typical, and effective, nudge which a reader will recall from previous chapters. Yet, the current discussion considers those who choose to *not* become donors, which is to say, to *not follow the nudge*. As with the speed hump example, declining here does not result in persistent prompts to reconsider until the "correct" (i.e., nudged) outcome is selected. Once a decision has been made, the decision-maker is free to continue about their day without facing the decision again and again.[17]

Now recall a second example, that of Google Maps. A GPS is typically considered a nudge insofar as it tells a person where to go but leaves the driving up to the driver (Sunstein, 2014; Thaler, 2018).[18] A GPS can also be seen as a hypernudge. For Yeung (2017, p. 122), the evidence for this comes from the fact that GPS, and Google Maps *specifically*, will automatically, in real-time,

> dynamically [reconfigure] the user's informational choice context … [T]he driver using Google Maps [is not] compelled to follow the "suggestions" it offers. But if the driver fails to follow a suggested direction, Google Maps simply reconfigures its guidance relative to the vehicle's new location.

In other words, and in contrast to the speed hump and opt-out organ donation, when the driver optsout of going left and instead chooses to go right, the driver will immediately be prompted with a nudge by the GPS to "do a U-turn," or some other "course-correcting" manoeuvre. Real-time feedback on the driver's location is integrated, the choice architecture is personalised, and the hypernudging GPS *immediately* responds to a person *not following the nudge* by nudging them again, and will *continuously* do so until the "correct" behaviour is exhibited. In short: hypernudges *follow*.

Of course, the decision-maker could opt-out of the hypernudge entirely – for instance, they could switch Google Maps off. But, as Lanzing (2019, p. 555) notes, this kind of extreme withdrawal is the *only* means of opting-out of hypernudges: "hypernudges cannot be opted out from without quitting this service altogether." This is to say nothing of the *cost* of opting-out itself. For instance, a driver who is using Google Maps to travel in a city they do not know, but deviates from the suggested course because of the obvious (at least to a human) unsuitability of the route (e.g., traffic, unsuitable roads), may find themselves frustrated by the constant nudges to return to the path they have deemed unsuitable, but an autonomous choice architect has deemed *probabilistically* the best route. Here, to switch off the hypernudge – to *opt-out* – is quite costly, because this driver is in a city which is alien to them.

This is, in part, a digital economy critique (Frischmann and Selinger, 2018; Mills, 2021), rather than an explicitly behavioural science criticism. But one may look to behavioural science to provide some commentary here. For instance, a hypernudging system which constantly follows a decision-maker imposes on their choices much more behavioural friction than a static nudge, and demands they must go against the choice architect multiple times, rather than once. Figure 4.1 provides an illustration of this.

If each decision to defy the nudge is taken to induce some cost to the decision-maker, the tendency for hypernudges to follow the decision-maker would lead to higher costs of going one's own way, which in turn would produce a

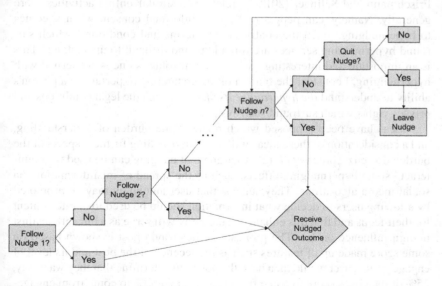

FIGURE 4.1 A hypernudging system

tendency for the decision-maker to follow the nudge (Frischmann and Selinger, 2018). Indeed, the option to quit the nudge only emerges after n iterations of nudging, presumably when the decision-maker is so exasperated at opting-out. In other words, the "default option" of a hypernudge is *always to nudge*.[19]

In sum, the burden of avoidance can be understood as the cost of opting-out of a hypernudge, given that hypernudges *follow* decision-makers because they can change *in response to* decision-makers. The example given here has been Google Maps GPS, but other examples support the notion of the burden of avoidance. The Facebook News Feed algorithm automatically curates content for users; the Google search algorithm automatically filters and orders results. Any deviation from these hypernudges is not met with force (e.g., "you *must* 'Like' this page") but rather immediate reconfiguration (e.g., "you didn't like that page? Well, here's another I'm sure you'll like instead"). Each avoidance of a variant of the nudge induces costs on decision-makers, while each reconfiguration in response to deviation demands those costs be incurred again.

The Burden of Understanding

Related to the burden of avoidance, but still separate, is what I call the burden of understanding. Yeung (2017) touches on the question of understanding in a discussion of informed consent, a topic which has also been discussed by Frischmann and Selinger (2018) in relation to similar online activities more generally. Namely, can people really give informed consent when it comes to hypernudging, given the contracts (e.g., terms and conditions) which surround hypernudging services are often long and difficult to understand? This is an important and interesting question, but insofar as one is concerned with hypernudging, I consider the burden of understanding to pertain to a person's ability to understand *the hypernudge itself*, rather than the legal conditions and privacy rights which surround such systems.

Ideas have been proposed which relate to the burden of understanding, and a consideration of these ideas will assist in revealing further aspects of the burden. Lorenz-Spreen *et al.* (2020) argue that nudging can be used to "counteract" some hypernudging effects, such as the spread of misinformation via social media algorithms. They suggest that user autonomy may be promoted by allowing users to decide what information should be used to curate content for their feeds and that this enhanced autonomy will serve as a bulwark against malign influence online. The proposal is thus: say a post is shown based on some score made up of features such as the recency of the post and its level of engagement; a user would then have the power to determine that they want, say, 25% of the post's score to come from recency, and 50% to come from engagement, and 25% to come from some other metric. Such a proposal would also be enhanced, Lorenz-Spreen *et al.* (2020) indicate, by widespread information disclosure revealing how a post's recency is determined or engagement calculated so that users can use this new autonomy in an informed way.

These measures *may*, if implemented, promote user autonomy, at least insofar as they grant users some control over the hypernudging AI to which they are subject. Yet, such a proposal raises two considerations. The first concerns the material interests of those who use hypernudges, such as social media companies. Lorenz-Spreen *et al.* (2020) seem to recognise the role of, say, proprietary rights in reducing user understanding, as noted in Chapter 1. But rather than viewing this as a challenge *independent* of understanding the hypernudge (i.e., the challenge of being *allowed to try to understand*) which requires a solution, they seem to assume a solution exists such that the solution to the challenge they are actually interested in (i.e., fostering understanding and autonomy) can be viable. This is not to denigrate individuals for trying to resolve problems, merely to recognise the role vested interests may play in obscuring well-meaning interventions.[20] The second consideration concerns the fact this intervention – essentially, a nudge – proposed by Lorenz-Spreen *et al.* (2020) is *in response to* hypernudging, rather than an explicit comment on hypernudging as a phenomenon. I will focus on this second consideration.

It is helpful to consider what a hypernudge such as the Facebook News Feed algorithm *actually* does. As noted in Chapter 3, on any given day, and for any typical user, the algorithm acts as an autonomous choice architect, selecting around 300 posts from a pool of around 1,500 posts to show the user, with the order being personalised based on some opaque criteria (Luckerson, 2015).[21] A 2014 investigation found that 75% of the variance in Facebook's News algorithm could be explained by only five variables, including total "Likes" and whether the post was made on a weekday or a weekend (Davies, 2014). However, in 2013, Lars Backstrom – the engineering manager for the News feed algorithm – claimed up to 100,000 variables were used.

In 2014, the true number of variables which would be relevant to consider probably lay somewhere in between these upper and lower bounds,[22] though as of writing, these estimates are likely rather inaccurate. For instance, in 2017 Facebook introduced "reactions," which allow users to express a range of reactions to a post, an evolution of the simple "Like" button. In 2018, it is reported that Facebook started prioritising "meaningful reaction," such as commenting and sharing over simple reactions such as liking, and as of 2020, Facebook has begun implementing credibility assessments when recommending content (Cooper, 2021). If I were to now ask the reader to write an equation for the Facebook algorithm, one may have an idea of the components involved but no certainty regarding what the AI is actually doing behind the scenes.

In some senses, this discussion has returned to Chapter 1. Hypernudges are, by their dynamic nature, often the product of AI systems with vastly more computational power than a human, which already places a burden on human understanding. Hypernudges, as are AI systems more generally, constantly changing, which further compounds confusion. But there is yet another layer to consider. One could describe the concept offered by Lorenz-Spreen *et al.* (2020) in simple mathematical terms:

$$Score = Ax + By + Cz \dots Nn \# (1)$$

Here, x, y, and z (and n, but this is just a placeholder) represent metrics established by the hypernudging choice architect (e.g., Facebook, Google, YouTube), such as the number of "Likes," the recency of the post, and the similarity of the post to content the decision-maker has previously engaged with. Furthermore, A, B, and C (and N; again, placeholder) represent percentage weights, such that $A + B + C$ (+N) = 100%. Decision-makers are, under the concept offered by Lorenz-Spreen *et al.* (2020), able to set the value of these weights.[23]

While this model appears quite simple[24] and the role of the weights quite understandable, this does not necessarily ensure that the *hypernudge* is understandable. For instance, how is a metric such as "the similarity of the post to the content the decision-maker has previously engaged with" determined? Alternatively, how can the recency of the post be determined; recent, *relative to what*? Finally, if the number of Likes is influenced by how others interact with the hypernudge, does this not mean that the choice architect and other decision-makers can influence what the decision-maker in question sees through their choices on underlying metrics (Sætra, 2020; Viljoen, 2020)?[25] This, of course, is all assuming that the score really is the sum of only three weighted variables. As the brief history of the Facebook algorithm suggests, the number of variables is likely much larger, and one cannot assume the relationship between variables is necessarily linear.

There is also a *control* aspect to this discussion, as Zuboff (2019) notes. For instance, if 100 variables are used to generate a score, and only five can be adjusted by the decision-maker, why only five, and why *those five in particular*? Even when trying to promote autonomy in hypernudged environments, it is likely a large degree of control needs to be retained by the choice architect simply because of the underlying complexity of the entire system (Delacroix and Lawrence, 2019). This, once more, is a point about motive power: if an AI is used because its materially superior motive power is necessary for computation, one cannot necessarily assume a human can replicate this computation. The best, the control perspective suggests, is that a human can be given control *of the wider service itself* rather than the systems (i.e., AI) which constitute it.

All of these factors inhibit understanding and produce a burden of understanding which raises the costs of autonomous decision-making; in such an environment, it is easier to just follow the nudge (Frischmann and Selinger,

2018). After all, ignoring the hypernudge often supposes one knows better; but if one does not know, or struggles *to know*, why an autonomous choice architect is "advocating" the options that they are, going against the grain becomes something of a leap of faith.[26] In sum, the burden of understanding can be understood as the cost of making a choice, which requires one to understand *why* one is being nudged one way or another. The underlying complexity of hypernudges means that decision-makers may struggle to act autonomously, and as with the burden of avoidance, simply default into following the hypernudge.

The Burden of Experimentation

The final burden is what I call the burden of experimentation. Experimentation is an implicit but necessary component of hypernudging, and the artifacts of this component have been seen already. For instance, in the above discussion of the burden of understanding, a simple model consisting of three percentage weights and three metrics was proposed. Yet, there was no particular reason to assume that the weighted metrics should have been combined into a single score. Nor was there any reason to believe that the combination of those products should be additive; why not subtract the first product from the second, or multiply the second by the third? Determining the optimal model for a particular hypernudge task is the same, in many cases, as determining the optimal model for any task an AI system might be required to perform, namely proceeding through a process of incrementally improving trial-and-error (i.e., reinforcement learning).

The burden of experimentation is not wholly concerned with the technical underpinnings of hypernudges which sections of this book have begun to entertain. But the notion that AI systems must experiment with different approaches to a task to identify the optimal solution is key to the burden of experimentation, as such experimentation can be understood as a cost decision-makers must endure. For instance, one may not know *a priori* what data should be collected in order to personalise a nudge (Sunstein, 2012). It may, therefore, be wise to collect a tremendous amount of data simply so that experiments can be conducted to *acquire* the knowledge of how best to personalise. This uncertainty creates a kind of logic of data accumulation insofar as because data *may be useful or valuable in the future*, if it can be collected today, it should be (Srnicek, 2016; Zuboff, 2019).[27] Yet this also creates costs for decision-makers in the form of privacy and surveillance (Lanzing, 2019; Sunstein, 2012; Yeung, 2017; Zuboff, 2019). These costs are the sort which should be understood as part of the burden of experimentation.

The burden of experimentation, as described above in terms of surveillance, may be understood as what Sætra (2020, p. 3) dubs "proactive surveillance," insofar as "it involves [the] use of surveillance in order to uncover information and change the actions of individuals." An example of such proactive surveillance within a hypernudging space comes from Frischmann and Selinger (2018), who report that Google Maps GPS will purposely send drivers on sub-optimal routes in order to gather data on under-mapped roads. Another famous, or perhaps *infamous*, example of experimentation is Facebook's mood experiment, where the News Feed algorithm was changed in such a way as to alter the mood of users, under the retroactive justification of learning how to improve Facebook services (Booth, 2014; Kramer, Guillory, and Hancock, 2014). One can describe both of these incidents in terms of costs; the driver faces a slower and more uncertain journey, and the Facebook user an erratic and uncertain mood. A final, recent example is that of the social media platform TikTok, which will periodically show a user a video they may *not* like, in order to learn more about the user (Wall Street Journal, 2021).

Hypernudging systems would also be expected to experiment with choice architecture. Russell (2019) has argued that AI systems often learn to maximise objectives through experimentation with people's preferences rather than learning how to satisfy individual preferences maximally. Writing on algorithms used by social media companies, Russell (2019, p. 8, original emphasis) states:

> Typically, such algorithms are designed to maximise *click-through*, that is, the probability that the user clicks on presented items. The solution is simply to present items the user likes to click on, right? Wrong. The solution is to change user's preferences so that they become more predictable.

Such a result could be seen as a consequence of using an AI system to nudge rather than a human choice architect. As was seen in Chapter 3, while an AI may possess superior computational power to architect choices, these same systems have very abstract views of human behaviour, and by extension welfare, dignity, respect (and so on), namely a view which is necessarily quantifiable and communicable to a machine. Given such limitations, it may not be surprising that one result of experimentation is to change *human* preferences, rather than to adjust *computational* processes. This also reflects an important question of interests. One may hope that the interests of the decision-maker, and of those who control the hypernudging system, align. For instance, Facebook may want to maximise the click-through rate, and this may be achieved by showing users the most worthwhile content. In such a special instance, the prerogative of the AI (maximise click-through rate) seems to benefit the decision-maker. But where they are not aligned, the AI does not necessarily place the decision-makers interests above that of its master (Willis, 2020). Rather, the opposite is true; it is the interests of those who control the AI which are served by the

TABLE 4.2   Three Burdens

BURDEN	DETAIL
Burden of Avoidance	• The challenge for decision-makers to "go their own way" created by hypernudges. • As hypernudges change in real-time, and in response to immediate feedback, hypernudging systems immediately nudge a decision-maker again.
Burden of Understanding	• The challenge for decision-makers to understand how and why they are being nudged, and to therefore make an informed decision. • As hypernudges are often proprietary, and as they often use significant amounts of data in abstract ways, a decision-maker may struggle to understand how they are being nudged, and simply defer to the nudge.
Burden of Experimentation	• The challenge for decision-makers to have their preferences respected. • As hypernudges need to learn how to nudge, decision-makers may often be subject to experimentation which will not respect their preferences.

AI, and decision-makers (and their interests) become interpreted as malleable factors within the AI calculation which can be changed to suit the calculation (Zuboff, 2019). Literally, *variables*.

Such experimental costs may compound. For instance, opposition to experimentation may be a reason why a decision-maker chooses not to follow the nudge; experimentation may be a fraction of the friction which constitutes the burden of avoidance. Equally, constant experimentation raises the costs of understanding. These three burdens are summarised in Table 4.2.

SUMMARY: CHAPTER 4

- AI systems allow human behaviours to be predicted more accurately, and choice architecture to be changed more rapidly, than a human choice architect ever could. This leads to the possibility of systems of nudges being used to subtly adjust (or control) preferences. These are hypernudges.
- Hypernudges are systems of personalised nudges. While each individual nudge may enable someone to "go their own way," hypernudges are more persistent and harder to escape.

- Hypernudges create several problems for human agency, and demand humans be experimented upon because of the need for the AI-driven hypernudge to learn.

NOTES

1. For instance, the explosion of misinformation through social media systems (van der Linden *et al.*, 2017), or the rise of internet addiction through extensively, behaviourally tuned programmes and applications (Aiken, 2017).
2. This perspective largely reflects Thaler and Sunstein's (2008) liberal economic perspective and could be challenged. See, for instance, Hausman and Welch (2010).
3. I do not mean this as an especially harsh comment towards the work of Karen Yeung (2017). Rather, I think the point of tension arises from a difference in discipline. Yeung (2017) has a legal background, and writes of hypernudging from this background, discussing principles, rights, obligations, and so forth. By contrast, the behavioural sciences operate – even if disputably – as a *science*, concerned with empirics and mechanisms. Of course, there is overlap, but in reflecting on the perspective I adopt and have adopted in the past within the discussion of hypernudging, I would argue these differences in perspective matter, but do not inherently point to criticisms or inadequacies in the work of others.
4. A reader should interpret Yeung's (2017) use of the word "potent" in the same way as potency has been discussed in this book.
5. A recent framework of a hypernudge has also been given by Morozovaite (2021). This framework, broadly, links the assumptions and doctrine of nudge theory to the three hypernudging elements identified by Lanzing (2019).
6. Several perspectives which I have and will express in this discussion, as well as motivations, are the product of personal correspondence with Henrik Sætra, to whom I am grateful.
7. As a rule-of-thumb, I will generally use the term "nudge" to refer to impersonal nudges, and specify when a nudge is personalised (i.e., "personalised nudge"). However, on occasion it may be worthwhile for the sake of emphasis or clarity to specify "impersonal nudge," as is the case here.
8. Indeed, a personalised nudging programme may determine that, for individuals who value choosing, no nudge should be used, and an active choice should be offered instead. I have taken to calling this quirk of personalisation "shielding," and may also be considered a kind of personalisation (Mills, 2022).
9. Incidentally, the speed hump example is fascinating for thinking about what a nudge is. For instance, the costs of ignoring the speed hump *may be* significant if it results in a collision or vehicle damage. Yet, these costs may *not* be if one assumes the chances of facing such costs are sufficiently low. An example: say the speed hump causes a person's vehicle damage with a cost of $10,000 to repair, but the chance of such damage occurring is 1 in 100,000. A nudge theorist may conclude that the implicit cost of the nudge (i.e., $0.10) is sufficiently small as to

not be significant (Thaler and Sunstein, 2008). This reveals the weakness of a definition built around significance. Another interesting perspective, revealed by the speed hump example, comes from noting *alternatives exist*. For instance, a sign simply asking a driver to slow down would also constitute a nudge under the common definition. If both a speed hump and a warning sign are nudges, one can clearly see it is possible to have nudges which exhibit different levels of *friction*, i.e., the discomfort of speeding over the hump is a kind of hedonic friction (i.e., it makes one feel uncomfortable), but there is no obvious hedonic friction from ignoring a sign on the side of the road.

10. See Chapters 2 and 3.

11. Another way of thinking about choice/delivery is thus: there are perhaps two dimensions along which people may differ, namely in preference, and in bias. Two individuals who differ in bias but not in preference likely respond to different types of nudges but benefit from being nudged towards the same outcome (i.e., delivery personalisation). Two individuals who differ in preference but not in bias likely respond to the same type of nudge but benefit from being nudged towards different outcomes (i.e., choice personalisation). Finally, two individuals who differ both in terms of bias and preference may benefit from being nudged towards different outcomes using different nudges.

12. Several recent studies demonstrate the validity of crude personalised nudges (Beshears *et al.*, 2021; Page, Castleman, and Meyer, 2020; Peer *et al.*, 2020; Schöning, Matt, and Hess, 2019). Page, Castleman, and Meyer (2020), for instance, personalise the reminder texts used to encourage high school students to complete their FAFSA application (a government programme designed to support students into higher education). These messages are personalised based on the progression of the individual student's application: a student who has not begun an application is sent a message to begin; a student who has completed part of the application is sent a message to finish; and a student who has finished is sent a message to make sure they have all the necessary additional materials prepared. Under the choice/delivery framework, Page, Castleman, and Meyer (2020) undertake personalised nudging insofar as the outcome individuals are nudged towards (i.e., begin, complete, review) is personalised based on application status (i.e., choice personalisation). Yet, this personalised nudge does not rely on big data. Much of the growing shift towards questions of heterogeneity, which inevitably lead to questions of personalisation, are arising due the decreasing "low hanging fruit" within behavioural science (Sanders, Snijders, and Hallsworth, 2018) and a push for more accurate investigation reliant on more complexity (Bryan, Tipton, and Yeager, 2021), not because of big data *per se*.

13. I will say slightly more on this all in Chapter 5, as thus far the notion of "feedback," has been taken very much as a given, but perhaps should not be.

14. Lanzing (2019, p. 555): "While nudges are also often not immediately detectable, they are and should be 'visible' in the physical world (we can 'see' the red arrow pointing to the staircase)."

15. See Frischmann and Selinger (2018, 2016) and Morozov (2013) for a similar discussion of how nudging and technology is increasingly pushing society towards a "frictionless future" (Morozov, 2013, p. x), and Bates (2020) for a discussion of how such a process marginalises human judgement and values.

16. I also worry this definition leaves an important aspect of hypernudging to be misunderstood. Aside from dynamism, an alternative definition of a hypernudge may be *nudges which are connected*. This is something of a literal interpretation of the term *hyper*nudge; just as *hyper*text describes the connections between bodies of text, and *hyper*space describes the space between planetary systems, so too might one see *hyper* as a description of the connection between nudges. Equally, this is also an interpretation reliant on a contemporary use of the word hyper. The original use of the word *hyper* was to describe something which was *over*, *above*, or *beyond* some normal state of being. In this sense, a *hyper*nudge could be interpreted as being an *extensive* nudge, or something *beyond* a nudge.

17. There are some limitations with these examples, which I argue would be the case for all examples. Reality rarely plays kindly with theory. For instance, most roads with a speed hump have multiple speed humps. This is a pedantic point, but still worth acknowledgement. In addition, a person may be prompted to become an organ donor when they renew their licence, or during several other administratively "convenient" instances.

18. I choose to discuss the example of a GPS as a nudge given the prior use of such an example in the literature. Nevertheless, this is an objectionable example. For instance, the GPS does not determine both the route to a destination and the destination *itself*, while a choice architect chooses *how* to nudge and *what* to nudge towards (hence the two-component framework for personalisation discussed above). I am grateful to Henrik Sætra for this comment.

19. Also see Frischmann and Selinger (2018) on *techno-social engineering*.

20. This discussion has been covered in much more detail by several authors, such as Pasquale (2015), Yeung (2017), Zittrain (2014) and Zuboff (2015, 2019).

21. The figures and processes I describe likely are out of date, but once more, owing to the proprietary nature of the algorithm, one cannot know for certain. I do not believe the strict accuracy of the figures is necessary, provided they are broadly accurate, for the point being made to be valid.

22. I suspect closer to the lower, simply based on the law of diminishing returns (i.e., the addition of a random second variable to a random first is likely going to change the amount of variance explained more than the addition of a random 100,000th variable to 99,999 others, because variance has an upper limit).

23. Which may, in itself, be an unfair assumption. See Öhman and Aggarwal (2020).

24. Assuming, of course, a certain degree of numerical ability on the part of the decision-maker, and general digital literacy.

25. To an extent, this is a reflection of the cybernetic (i.e., connected and feedback orientated) nature of a hypernudging system such as a social media algorithm. In this discussion, such connection and feedback obscures understanding, though I do not wish to present this as wholly a bad thing. Bates (2020), for instance, argues that these mechanisms enable political discourse and social interaction to occur, and that without such mechanisms, socially important (and politically contingent) aspects of human relations can be reduced to technocracy. The classic example is politics and voter apathy. If one does not believe their vote matters, they cease to vote, and the political dimension, strangely, becomes almost *de-politicised*, defaulting into a technocracy (which is perhaps tolerated, until it fails to respond to challenges of the time).

26. My comments here are inspired by the work of Moon (2010), who notes within marketing that the people who tend to resist marketing ploys are those who possess expert knowledge in an area, and the confidence that comes with it. Most people are not experts and will defer their judgement to that of the expert's. A marketer's job, Moon (2010) argues, is to be that expert for the masses.
27. I will discuss this logic more in Chapter 5.

Some Concluding Discussions

5

STRANGE LOOPS AND STRANGE REALITIES

Sherry Turkle (2004 [1984]) argues many computer scientists coming of age in the 1980s were influenced by Douglas R. Hofstadter's (2000 [1979]) book *Gödel, Escher, Bach.* The book focuses on the concept of the "strange loop" (p. 10), which is the name Hofstadter (2000 [1979]) gives to phenomena which emerge through self-reference and repetition. Such a concept is important in computer science because of the technique of *recursion*; used in functions which reference themselves, as briefly discussed in Chapter 2. Hofstadter (2000 [1979]) argues strange loops represent an "Eternal Golden Braid" through human thought, and indeed, perhaps throughout *all* intelligent thought. The characters which constitute the book's title, the mathematician Gödel, the artist Escher, and the composer Bach, are all offered as examples of revolutionary minds whose work is the product of strange loops.[1] Later in the book, Hofstadter (2000 [1979]) begins to entertain the strange loop between the human mind and AI, recognising that in the process of building an AI, humans engage in a process of self-reference. But perhaps the most prominent strange loop within *Gödel, Escher, Bach* is the dialogues between the Tortoise and Achilles. Hofstadter (2000 [1979]) uses these dialogues to offer ideas which may initially be familiar to the reader, before launching into discussions which prompt the reader to reflect on what they previously thought they knew from the dialogue (and everyday life).[2]

The Tortoise and Achilles discuss (amongst other things) the paradoxes of the Greek philosopher Zeno. These characters are comically logical, and exist outside of time, so are able to test Zeno's paradoxes and debate them when they break.[3] Consider Zeno's most famous paradox, that of the race

DOI: 10.1201/9781003203315-6

between the Tortoise and Achilles. The Tortoise is obviously much slower than Achilles, so much so that Achilles believes he can win the race even when the Tortoise begins halfway down the track. Of course, by the time Achilles has run half of the track, the Tortoise has "run" slightly further, say a quarter of the track. Then, once Achilles has run one-quarter, the Tortoise has "run," slightly further once again, say one-eighth of the track. Zeno's paradox here is that Achilles will never catch the Tortoise, despite Achilles being faster than the Tortoise.[4]

Mathematically, this race can be written as an infinite series: $\frac{1}{2} + \frac{1}{4} + \frac{1}{8} + \ldots + \frac{1}{2^n}$, for all values $n > 0$. This series can be shown to converge on the value of 1; as one's experience of the world teaches, Achilles can indeed beat the Tortoise in the race. But this only occurs, within the mathematical system of logic, when the series is allowed to go to infinity (i.e., $n \rightarrow \infty$). The series does not equal 1 if one stops counting at any point. Zeno's paradox arises, in part, because Zeno prompts one to stop counting, first at the halfway point, then three-quarters into the race, then seven-eighths, and so on.

There is much that could be said of Zeno's ideas, but this book is not the forum for this discussion, and I am not the writer to offer such a discussion. I bring Zeno and *Gödel, Escher, Bach* up for two reasons. Firstly, Zeno shows the potential beauty, but important limitations, of theoretical and logical thinking in a world which may be neither satisfied by theory nor consistent with logic (i.e., the *real* world). Secondly, because this chapter will in many ways be more about the people behind this book's titular subjects; the *people* behind AI, and the *people* behind behavioural science. Given this, I think it is important to at least keep as an artefact within one's mind – if not to directly integrate into the discussion – the works which (at least following Turkle (2004 [1984])) inspired the thinkers who built many of the systems and theories discussed in this book (Kuhn, 2012 [1962]).

This chapter initially focuses on three ideas which have close relations with one another and which emerge as points of discussion from the preceding chapters. These ideas are *distance from action, assumption of error,* and *behavioural logic*. In Chapter 3, the autonomous choice architect was introduced, and this in turn raised questions about the purpose and identity of human choice architects in a world inhabited by AI systems. The concept of *distance from action* will address these queries. In Chapter 4, I explained how autonomous choice architects can enable hypernudging and identified three burdens experienced by decision-makers, but I did not necessarily explain *why* such burdens seem acceptable to (behavioural) technologists developing them today. *Assumption of error* and *behavioural logic* will offer recourse to this point. To conclude this chapter, and this book, I will then give some thoughts

on the future of AI in behavioural science, centring my thoughts on the possibilities, and dangers, of a new *behaviourism*.

DISTANCE FROM ACTION

A question which emerges from the discussion of autonomous choice architects is *who is responsible for these entities*? *Who is responsible for an AI
which automatically architects choices to influence behaviour*? For Matthias
(2004), questions such as *who is responsible*? can only be answered by exam-
ining who is in control. A driver who chooses to drive through a red light
is *responsible* for their action (and may be *held responsible* insofar as they
suffer consequences for their actions; Oshana, 2002; Riceour, 2007) because
they are in control of the car and their choices. A child who chooses not to do
their homework is responsible for that decision (and may also be *held responsible* through measures such as detention). Responsibility, in other words, has
a close relationship with actions, and more intimately still, *control over those
actions*.[5]

Such a perspective is interesting when one considers questions of *unpredictability*. Control, Matthias (2004) argues, and as such responsibility,
is dependent on a person's ability to predict the results of their actions. For
instance, a driver who runs a red light, but who broke to *try and slow down*,
would probably receive more sympathetic responses from their peers. This is
to say, a driver would predict that braking would stop the car at the red light,
and if this action was taken, but the car did not stop, the responsibility for not
stopping is the fault of the brakes (and perhaps, subsequently, the car manufac-
turer) rather than the driver. Where systems are unpredictable, and thus where
it is harder to determine who is in control, Matthias (2004, p. 175) argues a
"responsibility gap" can emerge.

AI systems create such unpredictable environments, by virtue of their
ability to change, and the difficulty in understanding what AI is doing at any
one time. As such, as Gunkel (2017, pp. 311–312) argues, the emergence of
AI systems such as autonomous choice architects creates responsibility gaps
which require some redress:

> [M]achine learning systems, like AlphaGo,[6] are intentionally designed to
> do things that their programmers cannot anticipate, completely control, or
> answer for. In other words, we now have autonomous (or at least semi-auton-
> omous) computer systems that in one way or another have a "mind of their

own." And this is where things get interesting, especially when it comes to questions of responsibility.

If one accepts such a line of thinking, one may find themselves at an impasse. If, as Gunkel (2017) suggests, there may be a good reason to contend that the programmers of an AI system cannot predict the actions of the AI system, and – as Matthias (2004) argues – in such a circumstance there may be grounds to not hold these programmers *responsible*, who *should be held responsible*? Must one conclude, as Matthias (2004) muses is sometimes the case, that *no one is responsible*?[7] Perhaps an alternative perspective on the question of responsibility is required. Rather than beginning with the question of responsibility, I will try beginning with an example of nudging.

A classic example of architecting choices is changing the layout and position of food items on menus to encourage different meal choices (Dayan and Bar-Hillel, 2011; Krpan and Houtsma, 2020). One could imagine a human choice architect manually changing these menus or offering recommendations. This might be as simple as a waiter removing lunch menus at 14:00 PM, replacing them with evening menus, or it might be as complex and interpersonal as the waiter discussing recommendations, chef specialities, and wine pairings with customers. For the sake of discussion, I will label such activities *manual nudging*.

Yet, in previous chapters it was noted that such manual nudging can create inefficiencies. In the first instance, perhaps a customer would prefer the lunch menu to the evening menu, and that making it harder for a customer arriving at 14:15 PM to order is suboptimal. In the second instance, perhaps these personal recommendations from the waiter do enhance customer experiences and lead to better choices, but the waiter may not have enough time to provide this experience to every customer. In order to both reduce workload and improve the effectiveness of the choice architecture, a restaurant may turn to an automated choice architect. In the case of a restaurant menu, the restaurant could use an app to track customer purchases and offer personalised recommendations (Matz *et al.*, 2017), while an in-app menu could be used to learn customer biases and design personalised menus which suit the customer (Hauser *et al.*, 2009). The autonomous choice architect may not even be this advanced; simply by knowing the time of day, perhaps the AI system learns the optimal time to change the menus to suit the changing clientele (e.g., 14:37 PM, rather than 14:00 PM; Villanova *et al.*, 2021). I will label such activities *automatic nudging*.

Where does control rest in manual and automatic nudging? In manual nudging, much control rests with the waiter. For instance, if the waiter does not change the menus at the correct time, the responsibility for this rests with them (assuming the impact of this change is broadly what would be expected to occur) because they are in control of their actions. Equally, if a waiter is asked to recommend a wine pairing, and they suggest a pairing which would be expected to be very poor, the responsibility likely also rests on the waiter's shoulders, again because of their ability to control their recommendations. In automatic nudging, control *seems* to rest on the AI system. For instance, an autonomous choice architect which automatically analyses customer data using an unsupervised deep learning algorithm to identify optimal choice architecture is not being interfered with by a human. It is *unsupervised*, after all. Equally, no human is checking whether the menu design or recommendation is "optimal" for each customer – this process is automatic, and automated *precisely because humans cannot check*. Finally, no one – not the customer, not the company, and not the developer of the AI system – can necessarily explain why customer X receives menu Y or recommendation Z, and so one might argue that responsibility *cannot rest with anyone but the AI system* (Willis, 2020).

It would seem the responsibility gap has returned. Yet, the picture is not wholly complete. For instance, the above scenario is written as if the AI *knows what is optimal*, but this is not the case; AI systems need to be *told what is optimal* (Russell, 2019, 1997; Willis, 2020). Furthermore, the above scenario assumes the AI can do whatever it wants, but this is likely not the case. It would be silly, for instance, to allow the AI to show the breakfast menu at 20:00 PM or the evening menu at 08:00 AM. A well-designed autonomous choice architect will likely have fixed parameters under which it can operate, just as a well-briefed human choice architect will likely have fixed parameters under which they can operate (Thaler, 2021). Indeed, the optimisation function is a fixed parameter the AI cannot change. These parameters are not fixed for the waiter. Certainly, a waiter may want to keep their job, but this is an incentive the waiter has agency to reject. For instance, no matter what their boss tells them, there is nothing (in principle) stopping the waiter from changing the menus, or not changing the menus, or recommending to a customer a cocktail of red wine and Iron-Bru.

In short, an AI cannot have control of its actions because it does not have the agency *to act*; a waiter can have control of their actions because they have the agency *to act*.[8] As such, the former cannot bear responsibility, while the latter can (Sætra, 2021a). One may, then, begin to escape the responsibility gap which AI systems *appear* to create. For instance, by asking *who controls the AI parameters?* or even *who decided to use an AI system?*, it may be possible to determine that, say, the owner of the restaurant is responsible for the autonomous choice architect or the chief technology officer who gave the programmers the optimisation function (e.g., maximise profits), or the common sense programmer who decided the AI should not show the breakfast menu in the evening.[9]

I contend this discussion alone is valuable and should be recognised as extending the role of the human choice architect who is seeking to employ AI within their behavioural science practice to that of both an ethicist and data scientist. But I would also contend this discussion does not entirely address some dynamics of the discussion of responsibility. By focusing on agency, as opposed, merely, on motive power, one may be able to puncture the so-called "veil of complexity" (Sætra, 2021a, p. 87) argument, namely that responsibility is avoided because one cannot control that which cannot be predicted. Yet, such a puncture says little regarding *distance*, nor regarding *how people think about AI*.

By distance, I mean the number of entities between a person who controls an action, and the person who suffers (or prospers) from the action. As always, an example is useful. Consider a human choice architect tasked with designing a form to encourage, say, organ donation. The human choice architect knows people, on average, tend to choose whatever option is set as the default option, and so the form is designed such that people are organ donors, unless they choose otherwise (i.e., the default is changed; Johnson and Goldstein, 2003). The "distance" here is quite small: the choice architect designs the form, which the decision-maker receives and experiences.[10] Now consider this same task is done by an autonomous choice architect. Firstly, a human must design the AI and its parameters. This may involve just one entity (say, the human choice architect) or multiple entities (say, the human choice architect, *and* a programmer who writes the AI's code). Then, the AI generates the "optimal" choice architecture for the form. Assuming once the form is designed it is then distributed, the form is then experienced by the decision-maker. The "distance," in this instance, therefore, has grown, with the human choice architect separated from the decision-maker by an autonomous choice architect (at least).

In everyday life, people are familiar with this idea of distance. Politicians and CEOs will endlessly make claims to be "taking responsibility," because "the buck" (whatever that is) stops with them (Beer, 1993 [1974]). Another phrase which comes to mind is that adage, "a team is only as strong as its weakest link," implying that it is not the weakest link who is directly responsible, but a team who allows such weakness to go unsupported. The notion of responsibility flowing up some chain of command is perhaps idealistic. In reality, distance makes it harder to determine who is responsible and thus makes it more likely that the conclusion that *no one* is responsible is reached (Matthias, 2004). The addition of an AI system, even if it does not have agency to exceed its parameters, may increase the distance between the people in control of what is done (i.e., the action) and the people who experience what is done, allowing the former more opportunity to avoid reasonability.

Such a phenomenon likely occurs because of how people *think* about AI systems. Two examples have already occurred, and it is worth taking some time to consider them. In the first instance, a reader may have initially been stumped, or perhaps *convinced*, by the notion of a responsibility gap shielding humans from responsibility. Others may have been unconvinced, but I suspect even if one reached the conclusion that humans were reasonable, the reasoning was something like, "a person should be reasonable for implementing something they do not understand, because this is wilful ignorance." I do not dispute statements such as this but will point out that such a statement still implies that a human is responsible *not for the actions of the AI* but *for their own ignorance*. Merely by introducing an AI into the debate about responsibility, I would argue, one is liable to slip into arguments which are tangential to the initial question.

In the second instance, a reader may have found themselves, within the preceding chapters, confusing humans with machines and machines with humans. Indeed, while writing this book, I have found myself writing phrases which would seem to treat an AI as if it were the human; phrases like "AI systems *think*," "AI systems *understand*," or "AI systems *study*." But AI systems, as far as they exist today, do not really think, nor do they really understand or study. AI systems, insofar as one is limited by language, are best described as *calculating* or *analysing* (Watson, 2019). This is to say, it is perilously easy in everyday thinking to anthropomorphise and personalise AI systems, as well as to neutralise them, especially when they seem to work (Bates, 2020).[11] As such, one may fall into a trap of seeing an AI *as a valid target to bear responsibility*, because – by virtue of anthropomorphising – AI is given a kind of quasi-agency. The reverse may also be true; by neutralising and deferring to an AI, the AI may assume a kind of quasi-agency by virtue of a human's *rejection* of agency (Bates, 2020).

ASSUMPTION OF ERROR

The *assumption of error* is best summarised as the assumption that all decisions which a person could make contain some selection error which, if removed, would produce greater benefit for the decision-maker.

The problem of heterogeneity, discussed in Chapter 4, illustrates the assumption of error quite well. Consider an impersonal nudge which results in 70% of the population following the nudge, and 30% of the population not following the nudge. This nudge, in other words, has a potency of 70%, which may be considered quite effective. Yet, the problem of heterogeneity suggests that some of the 30% who did not follow the nudge would be *more likely* to follow the nudge if nudged differently, say via personalised nudging, because of their individual differences.[12] Assume, then, that personalised nudges are used, and this results in 90% of people choosing to follow the nudge, and 10% choosing not to follow the nudge. This is a clear, possibly significant, improvement in the nudge's potency (and, assuming the nudge confers welfare benefits, the average welfare of decision-makers; Peer *et al.*, 2020; Sunstein, 2012).

However, even with a 90% potency, the same rationale, stemming from the problem of heterogeneity, emerges: some within the 10% block will, owing to their individual differences, be more likely to follow the nudge if nudged differently.[13] Assuming sufficient data, and assuming sufficient value from doing so, the whole process of personalisation may go again, perhaps resulting in 99% following the nudge, and 1% not following. In fact, setting aside cost and data for a moment, this rationale can be repeated endlessly, and it does so because of a simple assumption made by choice architect: *decision-makers are always making some kind of error in judgement.* After all, if option A is determined to be the "best" option for a decision-maker (whatever "best" means), choosing option B is clearly an error in judgement. This is the assumption of error.

This assumption is grounded, in part, within nudge theory. For instance, implementing a nudge *before* any individual has even seen the choice which they must make assumes they will, on average at least, make some error in their judgement (Thaler and Sunstein, 2003). The assumption is also an expression of AI research. One could adopt quite a surface-level approach here, noting, for instance, that the very act of programming an AI to constantly reduce an error term implies that there is *always an error to be reduced* (Russell, 2019). But one could also take something of an ethnographic view of matters. For instance, Turkle (2004 [1984]) reports what might be described as early expressions of the assumption of error in her work on the development of the field of AI:

> The excursions into psychology and linguistics began as raids to acquire ideas that might be useful for building thinking machines. But the politics of "colonization" soon takes on a life of its own. The invaders come not only to carry off natural resources but to *replace native "superstitions" with their "superior" world view.* AI first declared the need for psychological theories that would work on machines. *The next step was to see these alternatives as better – better because they can be "implemented," better because they are more "scientific."*

(Turkle, 2004 [1984], pp. 229–230, emphasis added).

For Turkle (2004 [1984]), the early dialectic between the human mind and behaviour, on the one hand, and computational and logic systems, on the other, results in a "colonization" of human ideas (e.g., decision-making, intuition) with that of the machine, or the algorithm *within* the machine. While several syntheses have been proposed,[14] Turkle's (2004 [1984]) ethnographic research of early computer users points to a synthesis that elevates the computer's determinations because they are "more scientific" and, thus assumed, "better."[15] Such an assumption demands another – that the humans are *worse*, with this worseness shown by any deviation from the "better" determination. This necessary second assumption is the assumption of error.

It is not only Turkle (2004 [1984]) on which this proposition relies. Bates (2020) argues that, in the interaction between people and technology, people are liable to bend to the will of technology, as technologies are mechanistic and unyielding, while the human psyche has "plasticity" (p. 110) and can be moulded over time. Such an argument may still stand even when one considers AI systems can learn and change, because – as above – AI systems can learn and change within defined parameters, and it is these parameters which an AI cannot (currently) exceed, while a human can. Bates (2020) argues further that part of this moulding is simply a matter of convenience; that over time it is simply *easier* to follow algorithmic nudges (Morozov, 2013), an argument Frischmann and Selinger (2018) also offer, and an argument which clearly resonates with the three burdens discussed in Chapter 4. These ideas return one to the conflict of contradictions described above; in designing an artificial mind, one must contend with the material differences between the (flexible) human mind and the (inflexible) mechanical mind, and such a contention leads to a conclusion that it is the human mind which possesses flaws, and the human mind which should change to accommodate the mechanical.

One could describe this as a hierarchy, where the machine is placed above the human, and because the human mind will never ascend the hierarchy to match the machine (so the story may go), it must always be *in error* in comparison to the machine – in short, the assumption of error. As Bates (2020, p. 111) writes, "data-tracking and machine learning algorithms ceaselessly learn to 'solve' the key problems of modern life," assuming, of course, that those things which are "solved," such as judgements and preferences, are in fact problems (Broussard, 2018).[16] This process, Bates (2020, p. 111, original emphasis) continues, "increasingly eliminates human decision, and hence the possibility of *failure*, because the algorithms incorporate all deviation and exception into a ceaseless production of ever-new predictive models, in real time." Such a phenomenon could be described using a familiar term: *hypernudging*.[17]

BEHAVIOURAL LOGIC

The assumption of error leads to two important principles which constitute what I call *behavioural logic*. The first is that individuals should always be nudged. If all human decisions are assumed to contain some welfare-reducing error, assuming the benefits of nudging outweigh the costs (Sunstein, 2013, 2014, 2020b), the assumption of error means that there is always a *rationale* for nudging. The flipside of this may be to contend there is always a means of improving interventions (e.g., reducing an error term), often due to the emergence of new knowledge which one has now become privy to (Sætra and Mills, 2021). This is still the assumption of error, built upon a hierarchy of sorts: the decision-maker's ignorance is their error (or bias), while the new knowledge comes from insights discovered by the choice architect, increasingly via machine learning and big data (Mills and Sætra, 2022; Sætra, 2019; Yeung, 2017). Indeed, recent work has adopted this stance, be it from the perspective of decisional noise confusing human judgements (Kahneman, Sibony, and Sunstein, 2021), or from a neuro-philosophical perspective which questions the existence of free-will, giving licence to notions of algorithmic governance (Harari, 2015).[18]

Adopting a perspective of *improvement* is useful because it reveals the second principle of behavioural logic stemming from the assumption of error. Namely, that in order to improve, error must be *identifiable*. Within "traditional" nudging, errors in human decision-making are identified in the form of experiments, often adopting rigorous standards to isolate sources of variance (Della Vigna and Linos, 2022). For nudges which rely on algorithms and machine learning techniques (Benartzi, 2017; Weinmann, Schneider, and vom Brocke, 2016; Yeung, 2017), these autonomous choice architects must have data from which to learn (Mills and Sætra, 2022) and, as Watson (2019) notes, likely a *vast* amount of data. However, the assumption of error does not posit the presence of some specific decisional error *a priori*. Rather, it describes the belief that some error exists *somewhere* in the decision-making process, but without prior knowledge or expectation of *where* that error may be found. Therefore, the logical tendency which emerges, and the second principle of behavioural logic, is to capture *all behavioural data speculatively* (Srnicek, 2016; Zuboff, 2015, 2019)[19] and design interventions with the expectation that that behavioural data can yield worthwhile insights.[20]

Behavioural logic is a helpful tool in contrasting several ideas discussed in this book, and particularly Chapter 4. The belief that a person should always be nudged, premised on the belief there is always an error, provides a rationale for the introduction of automated choice architects, the design of hypernudging systems, and the introduction of AI systems into behavioural science more generally.[21] The search for a decisional error without a specific expectation of error also leads to AI systems designed with thousands of inputs and opaque computational designs, in part because *the programmers may not know what they are looking for*, or *what they are designing the machine to do specifically.* Finally, the premise of looking for an error without knowing where to look leads to practices such as encouraging people to share and generate data, while simultaneously attempting to treat individuals as more predictable, datafied subjects.

It is interesting to note that all of the practices and perspectives discussed above are not contingent on the assumption of error or behavioural logic being *accurate* descriptions of reality. One may believe, for example, that a decisional error is always present, when actually a decision-maker is making no error at all. Furthermore, one may believe a piece of data, or a piece of choice architecture, are important when these entities are merely artefacts disguising something else which perhaps does not demand AI or behavioural science to solve.[22] What matters is *belief*, and these concepts simply provide a window into the behaviour and development of autonomous choice architects, and perhaps the digital economy at large, which is only dependent on choice architects *believing* the assumption of error and behavioural logic to be correct.[23]

But there may be several reasons to disbelieve concepts such as the assumption of error and, as a result, behavioural logic. For instance, many of the costs associated with the three burdens discussed in Chapter 4, or the collection of data and the use of personalisation discussed throughout this book, are non-economic costs, such as privacy, autonomy, understandability, universality, and consent. These costs are opposed to a quite unique set of economic costs which surround the development of AI systems: data are pure public goods, meaning the cost of copying data is zero, while various technologies for collecting and analysing data are fixed costs with near-zero marginal cost (Frischmann and Selinger, 2018; Fuchs, 2019). As such, the easy-to-quantify economic costs of techniques such as hypernudging are often near-zero and tend towards zero the more nudging is done, and data are collected. Thus, the apparent benefit of these techniques (which is presumed to drive potency, which in turn motivates the use of these techniques in the first instance) need only be quite small to appear, in comparison to the economic costs, quite large.[24] Yet, this justification *necessarily ignores* the harder-to-quantify social costs of these techniques.[25]

Another valuable perspective may come from reflecting on alternative perspectives to problems which have, generally throughout this book, been assumed to only be surmountable by AI systems. For instance, is there an alternative approach to the problem of heterogeneity beyond personalisation? One interesting study by Arulsamy and Delaney (2020) may suggest so.

Investigating the effect of auto-enrolment on savings behaviour between those with, and those without, a mental health disability in the UK, Arulsamy and Delaney (2020) find that prior to the introduction of this *impersonal* auto-enrolment nudge, individuals with a mental health disability were significantly less like to save for retirement than their non-disabled counterparts. Some explanations for this disparity are that it is a direct result of the mental health disability (e.g., inability to understand retirement saving) or that it is an indirect result of the mental health disability (e.g., care routines taking up time which might otherwise be spent planning for retirement). Yet, for the purposes of this discussion, explanations for the disparity matter less than the fact that a disparity was found to exist, suggesting that the presence of a mental health disability was an important individual difference amongst workers within the UK in regard to workplace pensions.

After the auto-enrolment nudge was introduced, however, Arulsamy and Delaney (2020) report *no participation rate difference* between those with and those without a mental health disability, while the whole population on average began saving more. This result does not wholly discredit the problem of heterogeneity, nor the resulting assumption of error and behavioural logic. For instance, Bourquinn, Cribb, and Emmerson (2020) find the same policy may have negatively impacted individuals with lower incomes, suggesting another dimension of individual difference where the impersonal nudge may not have been optimal. Yet, the study by Arulsamy and Delaney (2020) does suggest that an abstract, logical approach to ideas such as heterogeneity, personalisation, potency, and indeed, behaviour itself may ignore important nuances about reality and may lead to highly logical *belief systems* which are not always suitable for either the task at hand or the values of the society in which they are expressed. Assumption of error and behavioural logic are summarised in Table 5.1.

TABLE 5.1   Assumption of Error and Behavioural Logic

PRINCIPLE	DETAIL
Assumption of Error	• All decisions contain mistakes which can be identified as decisional errors. • Human decisions are necessarily worse than the decision made by a well-specified algorithm.
Behavioural Logic	• People should always be nudged and nudging a person should be the default. • Because of the assumption of error, data should always be collected, with *as much data as possible* being collected, to identify the decisional error.

THE RETURN OF BEHAVIOURISM?

In conversations with colleagues over the past few years regarding the ideas contained within this book, a recurrent comment has involved *behaviourism*.[26] Such comments, I contend, are well founded, and it may be informative to spend some time digging into the role of behaviourism in relation to some of the ideas contained within this book.

Behaviourism is a psychological school of thought which holds:

1. Psychology is the science of behaviour, and that the mind is subordinate – as a subject of study – to behaviour.
2. Behaviour can be described and explained with reference *only* to external stimuli (e.g., lights, sounds, colours), and the mind (e.g., emotions, sensations) is merely a vehicle for expression rather than a contributory factor of expression.
3. All explanations of behaviour which use mental terms and *internal* psychology (i.e., the mind) can be restated in terms of external stimuli.

In short, behaviourism holds that human (and animal) behaviour is a function of external stimuli and that internal, neurological events may be ignored without losing insight into the behaviour of the target subject.

Various parallels exist between contemporary AI in behavioural science, as I have described it in this book, and behaviourism.[27] For instance, a foundational thinker in the world of behaviourism, John Watson, argued that the goal of behaviourism is "prediction and control" (Watson, 1913, p. 158), much in the same way processes such as hypernudging have been described in terms of predicting optimal choice architectures, and have, in turn, been accused of assailing human control (Smith and de Villiers-Botha, 2021).[28] Burrhus Skinner, the foundational thinker in so-called *radical* behaviourism, argued across various works (Skinner, 1984, 1976 [1974], 1953) that human behaviour was the product of environmental stimuli and could be shaped by manipulation of these stimuli with little attention paid to the internal environment. Such a perspective revels in ideas such as reinforcement learning, as well as in notions such as nudging behaviour, both of which have been seen prominently throughout this book.[29] Indeed, there are examples of Skinner's writing (e.g., Skinner, 1971) which seem to endorse systems such as autonomous choice architects and hypernudges within a social organisation.[30]

A consideration of the three points above reveals more parallels. Point (1) seems to demand behaviour can be observed and recorded as data, which follows both from the definition of behaviour given in Chapter 2, as well as the imperative to quantify feedback for AI systems noted throughout this book. Point (2) suggests the mind is merely a *site of computation*, receiving inputs and returning outputs, much in the same way one may view an AI system, but perhaps more crucially, *the way an AI system views human-beings*.[31] Finally, point (3) encourages one to draw equivalency between humans and machines, for any mental description can – according to behaviourism – be described in terms of external mechanisms; likewise, concepts such as behaviour, and definitions such as intelligence, have been established to speak in terms of *entities*, not human or computers (or animals). Two examples of this are the shift in behaviour from an *individual-level* view to a *population-level* view and the shift in intelligence from a multifaceted trait of a human (Sternberg, 1999) to a probabilistic perspective centred on behaviour.

I have chosen to highlight behaviourism here, rather than as an explicit philosophy of psychology throughout this book, because it speaks to a notion of a logical *belief system* established in the previous sections of this chapter. After all, computers and AI systems as commonly understood today are very much behaviourist creatures – *black boxes* – where some stimuli (i.e., *input*) produce some predictable behaviour (i.e., *output*).[32] Where human behaviour constitutes the AI system's inputs and outputs, it is easy to begin to see humans as a behaviourist creature too.

Yet, for all the equivalencies one may be able to draw, humans and computers are materially different. One of the key arguments, which set about the decline of behaviourism, came from Chomsky (1959), who noted that children possess a remarkable ability – once they have developed enough to *learn anything* – to learn languages very quickly. Chomsky (1959) noted that a behaviourist view of language learning would require a child to be bombarded with a language in order to learn it, but this is often not the reality under which children learn. The explanation, Chomsky (1959) argued, was that there are innate structures to language (and, one might argue, various other aspects of socialisation) which allow children to learn very rapidly and intuitively as the mind *naturally* develops in infancy. Such an argument constitutes a two-pronged attack on behaviourism. Firstly, the role of the external stimuli is greatly reduced. Secondly, the role of the internal environment (i.e., the mind) cannot be ignored.

If behaviourism is the psychological theory which underpins contemporary AI development, such a theory risks creating tensions – such as those intermated in Chapter 4 – when AI systems are used to influence human decision-making (Frischmann and Selinger, 2018). Behavioural logic and the assumption of error are, in several aspects, merely reflections of this behaviourist worldview. This is not to necessarily *attack* behaviourism as a perspective, and indeed, for one developing an AI, a behaviourist perspective may be very useful (McCarthy *et al.*, 1955; Simon, 1994 [1969]). It is merely to caution recognition of the possibility of tensions that come from the collision of disciplines, such as artificial intelligence and behavioural science. The cognitive revolution saw a turn away from behaviourism and more focus placed on empirical examination of the mind (Bechtel, Abrahamsen, and Graham, 1999; Miller, 2003; Turkle, 1988), while behaviourist-inspired perspectives such as reinforcement learning saw a radical shift in the development of AI systems (Russell, 1997; Simon, 1981). The use of artificial intelligence in behavioural science must reckon with this collision of paradigms and should proceed with recognition of the assumptions which – as a racing Tortoise and a Greek soldier might attest – do not always reconcile with reality.

SUMMARY: CHAPTER 5

- Autonomous choice architects raise important questions regarding responsibility. Because no human is directly nudging, it may be possible to not attribute responsibility to a human. However, because humans introduce and control autonomous choice architects, humans likely do retain responsibility.
- AI systems, when used to influence human behaviour, must make some strange assumptions and utilise some objectional logic. *Firstly*, humans must always be assumed to be making a decisional error. *Secondly*, humans should always be monitored to better train the AI system. These assumptions and logics may be criticised.
- Modern AI systems and contemporary behavioural science, when combined, seem to demonstrate features similar to the behaviourist school which was popular in the first half of the 20th century.

NOTES

1. Gödel's most famous contribution is the *incompleteness theorem*, which shows – by making mathematics self-referential – that not all true statements are provable. Escher's artistic style is famous for depicting physically impossible situations, such as staircases in loops, and tessellating shapes which, as one inspects the piece, seem to morph from one thing to another to another. Bach's compositional style was highly complex, with compositions formed from repetition of previous sections of the composition, often subtly different each time.
2. This is itself a strange loop.

3. As an aside, Hofstadter's (2000 [1979]) decision to offer the Tortoise and Achilles as characters wholly willing to accept logical propositions, but then test these propositions, offers a unique way of expressing the Gödel's incompleteness theorem. When a logical proposition seems to break, the characters are left in a quandary between a system of logic which is internally complete but, when expressed, is wholly inadequate.

4. There may have been more philosophical implications to this idea also. When the race is constructed in this way, neither competitor will ever finish the race. Thus, the paradox may also be interpreted as a reflection on the futility of trying to reach B from A. This is to say, because B will never be reached, a person should focus on their journey from A, rather than focusing on reaching their destination, B.

5. Matthias (2004) also notes there are often situations where *no one* can be considered directly responsible, owing to no individual having sufficient control. The example given is the piloting of a spacecraft on Mars. If a technician causes the craft to fall into a crater and become unusable, the technician may be blamed. But if this incident occurred because of unsuitable and *uncontrollable* weather conditions on Mars, or time delays between Earth and Mars, these factors which are outside of the technician's control often mean the technician would not be held responsible, but *neither would anyone else.*

6. Google's sophisticated AI system designed to play the game Go, and which beat the world Go champion in 2016.

7. For a recent, interesting contribution, consider the discussion offered by Willis (2020) on AI generated dark patterns. Dark patterns are choice architectures used in website design to manipulate users. Willis (2020) argues that dark patterns are often not the direct design of marketing executives, but may be the *accidental* result of marketing executives using AI systems to maximise some reward function (e.g., increase sales). These systems may subsequently learn that dark patterns maximise the reward function, and implement them, despite never *explicitly* being told to. This is somewhat a discussion for the debate regarding AI control. See Russell (2019).

8. It is interesting to reflect on this argument in relation to the discussion of motive power. From a material perspective, the motive power argument advances much of the discussion of AI within behavioural science, as the preceding chapters show. But there are clear limits, philosophically speaking, to the role of motive power within this field.

9. In my work with Henrik Sætra (i.e., Mills and Sætra, 2022), we offer five ways in which humans continue to influence AI systems. I must credit Sætra with these points:

   1. Humans programme the systems and algorithms involved, and this will always entail making a range of choices regarding how the system will end up making decisions.
   2. Humans decide how, where, and when these systems are to be applied.
   3. Humans directly instruct these systems to optimise based on selected variables.
   4. Humans are involved in deciding which variables or factors (or choice architectures) are manipulable by the AI system.

5. Humans are involved in a wide range of actions that shape and influence the generation, selection, and codification of the data used by these systems.

10. One might imagine some additional steps. For instance, perhaps the choice architect works for a private consultancy contracted by a government. Then, one might suggest the "distance" is consultant → government → decision-maker.

11. By "neutralise," I mean to forget that AI systems are value-laden entities which are the products of various human decisions and could be constructed differently depending on who is constructing them.

12. It also holds that some of those in the 70% who *did follow* the nudge would also become more likely to follow the nudge if nudged differently. This may seem counter-intuitive, but to an AI, there may be people who – under one construction of choice architecture – are predicted to have a 60% chance of following the nudge, and – under a different construction of choice architecture – are predicted to have a 90% chance of following the nudge. Probabilistically, this person follows the nudge under both constructions, but the "risk" of them not following is reduced in the second construction.

13. And, as with the previous footnote, some in the 90% group may also be determined to be *more likely* to follow the nudge if nudged differently.

14. See, for example, Brynjolfsson and McAfee (2014), Coeckelbergh (2020), Russell (2019) and Tegmark (2017).

15. Such language is reminiscent of that used in Thaler and Sunstein's (2003, p. 175) original paper on the concept of nudging. For instance: "In our understanding, a policy counts as 'paternalistic' if it is selected with the goal of influencing the choices of affected parties in a way that will make those parties better off. We intend 'better off' to be measured as objectively as possible, and we clearly do not always equate revealed preference with welfare. That is, we emphasize the possibility that in some cases individuals make inferior choices, choices that they would change if they had complete information, unlimited cognitive abilities, and no lack of willpower." Thaler and Sunstein (2008, p. 5) do subsequently develop this proposal to be, "better off, as judged by themselves," themselves being the decision-maker. This *perhaps* allows some criticism to be satisfied, but it must still be noted that if individuals knew what options would lead them to be better off (Selinger and Whyte, 2010), they would never choose a worse option. An external party (i.e., a choice architect) still has to try and determine what would leave a decision-maker better off, and I have no reason to believe Thaler and Sunstein (2008) would not advocate this be done "as objectively as possible." Furthermore, by allowing a decision-maker's revealed preference to be ignored because it may be "erroneous," there is perhaps no way of determining when a decision-makers judgement is "as judged by themselves" until it is that which matches the judgement of the external party (a variant on the burden of avoidance). Until such a match in "preferences," occurs, one may also find reason to keep nudging the decision-maker. At best, the "as judged by themselves," criterion simply acknowledges differences exist between individuals, and that some personalisation may ultimately be appropriate. It does little to counter the arguments for nudging, nor little to defend the judgement of individuals.

16. Also see Broussard (2018), who has dubbed the tendency to believe all problems have a technological solution *technochauvinism*.

17. Also see Frischmann and Selinger's (2018, p. 209) concept of "engineered determinism": "Engineered determinism occurs when techno-social environments control how people behave, develop, and relate to each other." Also consider Zuboff's (2019, p. 15) concept of "behavioural futures," or bets on future behaviour similar to a futures product in a financial market. My view is that the notion of *determinism* may go too far; human behaviour, at least to a machine, is *probabilistic*, and likely never so certain as to be determinable. I have expressed such a view to Frischmann in personal correspondence. Yet, I have also expressed my view that regardless of whether it is achievable, something akin to engineered determinism may *realistically* be the *ambition* of some designers of what Frischmann and Selinger (2018) call techno-social environments, or what I have here called autonomous choice architects.

18. Also see Turkle (2004 [1984], 1988) for a qualitative account of the development of these ideas of the mind as a computer programme in computing circles during the 1970s and 1980s.

19. Also see Srnicek (2016), who argues simply because technologists may not know what data will be useful in the future, but have an expectation that *some* data will be useful, choose to collect *all* possible data.

20. As a brief example, two methods of personalising nudges which I am familiar with are moderated regression analysis, and neural networks. The exact details of these methods are not important. What is important is that neither method determines whether personalisation is *possible*. Rather, these methods allow one to test whether a particular input variable x can be used to personalise choice architecture y. But if one is determined to personalise y, perhaps because the potency of y does not meet some threshold and the problem of heterogeneity is believed, then one can use these methods to test countless input variables until some variable, seemingly, produces the desired result. If one assumes these variables are, say, individual differences between people, then one may begin to find all sorts of "individual differences," on which to train models, from the obvious (e.g., gender, age, income level) to the less obvious (e.g., personality profile, Facebook "Likes") to the subconscious (e.g., pupil dilation, blood pressure). The field of, and technologies involved in, the collection of such behavioural data is known as *behavioural informatics* (Fu and Wu, 2018; Krpan and Urbanik, 2020).

21. As the pursuit of error demands greater computational demands.

22. For instance, from a statistical perspective, x may be used to explain y, and may do it quite well. But if the "driver," of x (i.e., that which causes x to vary) is actually z, one need not collect both x and z to explain y. By way of a behavioural science example, a recent article regarding the lack of uptake of life insurance by millennials documented several technology start-ups using AI and behavioural science to encourage this demographic to get insured through techniques such as gamification. The article concluded, however, by acknowledging that none of these strategies can overcome the lack of wealth and significant debt burden facing American millennials (Waters, 2021).

23. For an informative perspective on the digital economy of this matter, see Sadowski (2019). For a discussion of fallacious beliefs within the digital economy, see Morozov (2013).

24. It is worth taking a moment to consider the mathematics of this proposition. The equation $\frac{cost}{benefit}$ where *benefit* = 1 describes the cost of receiving 1 unit of benefit (i.e., cost per benefit), while the inverse, $\frac{benefit}{cost}$ where *cost* = 1 describes the benefit received from paying 1 unit of cost (i.e., benefit per cost). For my purposes here, the former (cost per benefit) equation is more useful. Cost can be taken to consist of two components, namely fixed costs and variable costs, such that $cost\ per\ benefit = \frac{(fixed + variable)}{benefit}$ Because, once a means of collecting data has been established (a fixed cost, e.g., the Facebook platform), the variable cost of collecting data is negligible (e.g., electricity costs, maintenance costs). Therefore, one can assume *variable* = 0 But if one holds that, say, more data produces more knowledge which produces more benefit (e.g., because one learns how to nudge better), benefit can increase as cost stays the same (i.e., fixed cost is constant, variable cost is zero), and so the cost per benefit tends towards zero as benefit tends towards infinity.

25. For instance, consider the so-called "privacy paradox" (Gerber, Gerber, and Volkamer, 2018, p. 226), where people report valuing their privacy much more than their subsequent behaviour would suggest. The privacy paradox raises a great-many questions, but also demonstrates that *accurately valuing privacy may be a difficult task*.

26. A useful account on the relationship between big data systems and behaviourism is given by Sætra (2018).

27. For an interesting, contrasting perspective, see Turkle (1988), who argues that the emergence of AI actually serves to discredit behaviourism as a field of study.

28. The above statement is taken from Watson's (1913) *Psychology as the Behaviorist Views It*, a paper which has also been dubbed the "Behaviourist Manifesto." The full quote (Watson, 1913, p. 158) is, "Psychology as the behaviorist views it is a purely objective experimental branch of natural science. Its theoretical goal is the prediction and control of behaviour. Introspection forms no essential part of its method, nor is the scientific value of its data dependent upon the readiness with which they lend themselves to interpretation in terms of consciousness. The behaviorist, in his efforts to get a unitary scheme of animal response, recognizes no dividing line between man and brute. The behavior of man, with all of its refinement and complexity, forms only a part of the behaviorist's total scheme of investigation." Watson, incidentally, would leave academia after the revelation of an affair, and would go on to a very successful career in marketing. A reader may make of this what they will.

29. It is also curious to note that Skinner's (1984) perspective is not that stimuli *x* will produce behaviour *y*, but that through the development of what Graham (2019) calls *learning histories*, an individual will become *more likely* to demonstrate behaviour *y* given stimuli *x*. In short, Skinner (1984) couches behaviour in a somewhat familiar probabilistic language.

30. Skinner (1971, p. 198): "[I]t is in the nature of an experimental analysis of human behavior that it should strip away the functions previously assigned to a free or autonomous person and transfer them one by one to the controlling environment."

31. Skinner (1976 [1974], p. 192), rather curiously, describes the human mind as a "black box," a term used in Chapter 1 to describe both AI systems and the human mind. In this instance Skinner (1976 [1974]) uses the term to describe something (i.e., the mind) which is not *beyond* understanding, but merely *unnecessary* to understand. This somewhat underscores the epistemology of behaviourism: if one can accurately predict an output given a specified input, behaviourists do not concern themselves with the transformative process occurring in between.

32. This perspective stands in contrast to the perspective adopted by Turkle (1988), namely that computers killed behaviourism by demonstrating that the "internal states" of machines mattered to the behaviour they demonstrated, and thus the "internal states" of humans probably mattered too. From this perspective, computers *cannot* be behaviourist creatures. However, I would conjecture that as computers have become more ubiquitous in everyday life, they have – in the philosophical sense – become more transparent; seen not as an *entity* with which to interact, but as a means to some other end (Susser, 2019). Thus, I would suggest the internal state of the computer which did so much to weaken behaviourism has, in contemporary society, become invisible, and allowed computers to *appear* as behaviourist creatures.

References

Adadi, A., Berrada, M. (2018) 'Peeking Inside the Black-Box: A Survey on Explainable Artificial Intelligence (XAI)' *IEEE*, 6, pp. 52142–52160

Aiken, M. (2017) *The Cyber Effect*. John Murray, UK

Akerlof, G. A. (1991) 'Procrastination and Obedience' *The American Economic Review*, 81(2), pp. 1–19

Aldohni, A. K. (2021) 'The Accessibility of Credit and the Protection of Consumers in the High-Costs Credit Sector: A Multifaceted Challenge' *Duke Law Journal*, 84, pp. 197–213

Allcott, H. (2011) 'Social Norms and Energy Conservation' *Journal of Public Economics*, 95, pp. 1082–1095

Allcott, H., Rogers, T. (2014) 'The Short-Run and Long-Run Effects of Behavioral Interventions: Experimental Evidence from Energy Conservation' *The American Economic Review*, 104(10), pp. 3003–3037

Altmann, S., Grunewald, A., Radburch, J. (2021) 'Interventions and Cognitive Spillovers' [Online] [Date accessed: 14/12/2021]: http://www.restud.com/wp-content/uploads/2021/11/MS26888manuscript.pdf

Aonghusa, P. M., Michie, S. (2021) 'Artificial Intelligence and Behavioral Science Through the Looking Glass: Challenges for Real-World Application' *Annals of Behavioural Medicine*, 54, pp. 942–947

Arulsamy, K., Delaney, L. (2020) 'The Impact of Automatic Enrolment on the Mental Health Gap in Pension Participation: Evidence from the UK' UCD Geary Institute for Public Policy Discussion Paper Series no. WP2020/04. [Online] [Date accessed: 08/02/2021]: https://www.ucd.ie/geary/static/publications/workingpapers/gearywp202004.pdf

Ashby, W. R. (1978) *Design for a Brain*.Chapman and Hall, UK

Bang, H. M., Shu, S. B., Weber, E. U. (2020) 'The Role of Perceived Effectiveness on the Acceptability of Choice Architecture' *Behavioural Public Policy*, 4(1), pp. 50–70

Bates, D. (2020) 'The Political Theology of Entropy: A Katechon for the Cybernetic Age' *History of the Human Sciences*, 33(1), pp. 109–127

Bechtel, W., Abrahamsen, A., Graham, G. (1999) 'The Life of Cognitive Science' in Bechtel, W., Graham, G. (eds.) *A Companion to Cognitive Science*, Blackwell, USA

Becker, K. (2017) 'When Computers Were Human: The Black Women Behind NASA's Success' *New Scientist*. [Online] [Date accessed: 23/03/2021]: https://www.newscientist.com/article/2118526-when-computers-were-human-the-black-women-behind-nasas-success/

Beer, S. (1979 [1966]) *Decision and Control: The meaning of Operational Research and Management Cybernetics* Wiley, UK

Beer, S. (1993 [1974]) *Designing Freedom*. Anansi, USA

Beggs, J (2016) 'Private-Sector Nudging: The Good, the Bad and the Uncertain' in Abdukadirov, S (eds.) *Nudge Theory in Action*, Palgrave Macmillan, London

Benartzi, S. (2017) *The Smarter Screen: Surprising Ways to Influence and Improve Online Behavior*. Portfolio, UK

Benartzi, S., Thaler, R. H. (1995) 'Myopic Loss Aversion and the Equity Premium Puzzle' *The Quarterly Journal of Economics*, 110(1), pp. 73–92

Bernheim, B. D. (1994) 'A Theory of Conformity' *Journal of Political Economy*, 102(5), pp. 841–877

Beshears, J., Dai, H., Milkman, K. L., Benartzi, S. (2021) 'Using Fresh Starts to Nudge Increased Retirement Savings' *Organizational Behavior and Human Decision Processes*, 167, pp. 72–87

Bleier, A., Harmeling, C. M., Palmatier, R. W. (2019) 'Creating Effective Online Customer Experiences' *Journal of Marketing*, 83(2), pp. 98–119

Booth, R. (2014) 'Facebook Reveals News Feed Experiment to Control Emotions' *The Guardian*. [Online] [Date accessed: 08/03/2021]: https://www.theguardian.com/technology/2014/jun/29/facebook-users-emotions-news-feeds

Bourquin, P., Cribb, J., Emmerson, C. (2020) 'Who Leaves Their Pension After Being Automatically Enrolled?' Institute for Fiscal Studies Briefing Note BN272. [Online] [Date accessed: 19/05/2020]: https://ifs.org.uk/uploads/Who-leaves-their-pension-after-being-automatically-enrolled-BN272.pdf

Bovens, L. (2008) 'The Ethics of Nudge' in Grüne-Yanoff, T., Hansson, S. O. (eds.) *Preference Change: Approaches from Philosophy*. Springer, New York

Broniarczyk, S. M., Griffin, J. G. (2014) 'Decision Difficulty in the Age of Consumer Empowerment' *Journal of Consumer Psychology*, 24, pp. 608–625

Broussard, M. (2018) *Artificial Unintelligence: How Computers Misunderstand the World*. MIT Press, USA

Brown, C. L., Krishna, A. (2004) 'The Skeptical Shopper: A Metacognitive Account for the Effects of Default Options on Choice' *Journal of Consumer Research*, 31(3), pp. 529–539

Bruns, H., Kantorowicz-Reznichenko, E., Klement, K., Jonsson, M. L., Rahali, B. (2018) 'Can Nudges be Transparent and Yet Effective?' *Journal of Economic Psychology*, 65, pp. 41–59

Bryan, C. J., Tipton, E., Yeager, D. S. (2021) 'Behavioural Science Is Unlikely to Change the World Without a Heterogeneity Revolution' *Nature Human Behaviour*. https://doi.org/10.1038/s41562-021-01143-3

Brynjolfsson, E., McAfee, A. (2014) *The Second Machine Age: Work, Progress, and Prosperity in a Time of Brilliant Technologies*. W. W. Norton, UK

Bucher, T. (2016) 'Neither Black Nor Box: Ways of Knowing Algorithms' in Kubitschko, S., Kaun, A. (eds.) *Innovative Methods in Media and Communication Research*. Palgrave Macmillan, USA

Buolamwini, J., Gebru, T. (2018) 'Gender Shades: Intersectional Accuracy Disparities in Commercial Gender Classification' *Proceedings of Machine Learning Research*, 81, pp. 1–15

Burrell, J. (2016) 'How the Machine 'Thinks': Understanding Opacity in Machine Learning Algorithms' *Big Data and Society*. https://doi.org/10.1177/2053951715622512

Chomsky, N. (1959) 'A Review of B. F. Skinner's Verbal Behavior' *Language*, 35(1), pp. 26–58

Chung, T. S., Wedel, M., Rust, R. T. (2016) 'Adaptive Personalization Using Social Networks' *Journal of the Academy of Marketing Science*, 44, pp. 66–87

Coeckelbergh, M. (2020) 'AI Ethics. MIT Press, Cambridge, MA

Cooper, P. (2021) 'How the Facebook Algorithm Works in 2021 and How to Make it Work for You' Hootsuite. [Online] [Date accessed: 08/03/2021]: https://blog.hootsuite.com/facebook-algorithm

Croxson, K., Feddersen, M., Burke, C. (2019) 'Robo Advice: Will Consumers Get With the Programme?' *FCA*. [Online] [Date accessed: 25/07/2021]: https://www.fca.org.uk/insight/robo-advice-%E2%80%93-will-consumers-get-programme

Danziger, S., Levav, J., Avnaim-Pesso, L. (2011) 'Extraneous Factors in Judicial Decisions' *Proceedings of the National Academy of Sciences of the United States of America*, 108(17), pp. 6889–6892

Darmody, A., Zwick, D. (2020) 'Manipulate to Empower: Hyper-relevance and the Contradictions of Marketing in the Age of Surveillance Capitalism' *Big Data and Society*, 7(1), pp. 1–12

Davies, S. (2014) 'We Cracked The Code On How The Facebook News Feed Algorithm Works' *The Federalist*. [Online] [Date accessed: 08/03/2021]: https://thefederalist.com/2014/02/20/we-cracked-the-code-on-how-the-facebook-news-feed-algorithm-works/

Dayan, E., Bar-Hillel, M. (2011) 'Nudge to Nobesity II: Menu Positions Influence Food Orders' *Judgment and Decision Making*, 6(4), pp. 333–342

de Bellis, E., Hildebrand, C., Ito, K., Herrmann, A., Schmitt, B. (2019) 'Personalizing the Customization Experience: A Matching Theory of Mass Customization Interfaces and Cultural Information Processing' *Journal of Marketing Research*, 56(6), pp. 1050–1065

de Vos, J. (2020) *The Digitalisation of (Inter)Subjectivity: A Psy-critique of the Digital Death Drive*. Routledge, UK

Delacroix, S., Lawrence, N. D. (2019) 'Bottom-up data Trusts: distributing the 'one size fits all' approach to data governance' *International Data Privacy Law*, 9(4), pp. 236–252

Della Vigna, S., Linos, E. (2020) 'RCTs to Scale: Comprehensive Evidence from Two Nudge Units' *Econometrica*, 90(1), pp. 81–116

Diemand-Yauman, C., Oppenheimer, D. M., Vaughan, E. B. (2011) 'Fortune favours the **bold** (*and the italicized*): Effects of disfluency' *Cognition*, 118(1), pp. 111–115

Dolan, P. (2019) *Happy Ever After: A Radical New Approach to Living Well*. Penguin Books, UK

Dolan, P., Galizzi, M. M. (2015) Like Ripples on a Pond: Behavioral Spillovers and Their Implications for Research and Policy' *Journal of Economic Psychology*, 47, pp. 1–16

Dolan, P., Hallsworth, M., Halpern, D., King, D., Metcalfe, R., Vlaev, I. (2012) 'Influencing behaviour: The mindspace way' *Journal of Economic Psychology*, 33(1), pp. 264–277

Dourish, P. (2016) 'Algorithms and Their Others: Algorithmic Culture in Context' *Big Data and Society*. https://doi.org/10.1177/2053951716665128

Duranty, J., Corbin, T. (2022) 'Artificial Intelligence and Work: A Critical Review of Recent Research from the Social Sciences' *AI and Society*. https://doi.org/10.1007/s00146-022-01496-x

Ellsberg, D. (1961) 'Risk, Ambiguity, and the Savage Axioms' *The Quarterly Journal of Economics*, 75(4), pp. 643–669

Ferwerda, B., Tkalcic, M. (2018) 'Predicting Users' Personality from Instagram Pictures: Using Virtual and/or Content Features?' UMAP'18. https://doi.org/10.1145/3209219.3209248

Floridi, L. (2020) 'The Fight for Digital Sovereignty: What It Is, and Why It Matters, Especially for the EU' *Philosophy and Technology*, 33, pp. 369–378

Forde, J. Z., Paganini, M. (2019) 'The Scientific Method in the Science of Machine Learning' ArXiv [Online] [Date accessed: 15/09/2021]: https://arxiv.org/abs/1904.10922

Fraser, C. (2020) 'Target Didn't Figure Out a Teenager Was Pregnant Before Her Father Did, and that One Article that Said They Did Was Silly and Bad' *Medium*. [Online] [Date accessed: 24/07/2021]: https://medium.com/@colin.fraser/target-didnt-figure-out-a-teen-girl-was-pregnant-before-her-father-did-a6be13b973a5

Friesen, N. (2010) 'Mind and Machine: Ethical and Epistemological Implications for Research' *AI and Society*, 25, pp. 83–92

Friestad, M., Wright, P. (1994) 'The Persuasion Knowledge Model: How People Cope with Persuasion Attempts' *Journal of Consumer Research*, 21(1), pp. 1–31

Friestad, M., Wright, P. (1999) 'Everyday Persuasion Knowledge' *Psychology and Marketing*, 16(2), pp. 185–194

Frischman, B., Selinger, E. (2016) 'Utopia?: A Technologically Determined World of Frictionless Transactions, Optimized Production, and Maximal Happiness' *UCLA Law Review Disclosure*, 64, pp. 372–391

Frischmann, B., Selinger, E. (2018) *Re-engineering Humanity*. Cambridge University Press, UK

Fu, Y., Wu, W. (2018) 'Behavioural Informatics for Improving Water Hygiene Practice Based on IoT Environment' *Journal of Biomedical Informatics*, 78, pp.156–166

Fuchs, C. (2019) *Rereading Marx in the Age of Digital Capitalism*. Pluto Books, UK

Furr, M. R. (2009) 'Personality Psychology as a Truly Behavioural Science' *European Journal of Personality*, 23, pp. 369–401

Garcez, A. D., Lamb, L. C. (2020) 'Neurosymbolic AI: The 3rd Wave' [Online] [Date accessed: 24/11/2021]: http://arxiv.org/abs/2012.05876v2

Garnelo, M., Shanahan, M. (2019) 'Reconciling Deep Learning with Symbolic Artificial Intelligence: Representing Objects and Relations' *Current Opinion in Behavioral Sciences*, 29, pp. 17–23

Gerber, N., Gerber, P., Volkamer, M. (2018) 'Explaining the Privacy Paradox: A Systematic Review of Literature Investigating Privacy Attitude and Behavior' *Computers and Security*, 77, pp. 226–261

Gigerenzer, G. (2007) *Gut Feelings: Short Cuts to Better Decision Making*. Penguin Books, UK

Gigerenzer, G. (2018) 'The Bias Bias in Behavioral Economics' *Review of Behavioral Economics*, 5, pp. 303–336

Golman, R., Loewenstein, G. (2018) 'Information Gaps: A Theory of Preferences Regarding Presence and Absence of Information' *Decision*, 5(3), pp. 143–164

González-Cabañas, J., Cuevas, A., Cuevas, R., López-Fernández, J., García, D. (2021) 'Unique on Facebook: Formulation and Evidence of (Nano)targeting Individual Users with non-PII Data' IMC'21. https://doi.org/10.1145/3487552/3487861

Graham, G. (2019) 'Behaviorism' in Zalta, E. N. (eds) *The Stanford Encyclopedia of Philosophy*. [Online] [Date accessed: 09/08/2021]: https://plato.stanford.edu/cgi -bin/encyclopedia/archinfo.cgi?entry=behaviorism

Gunkel, D. J. (2017) 'Mind the Gap: Responsible Robotics and the Problem of Responsibility' *Ethics and Information Technology*, 22, pp. 307–320

Haig, B. D. (2014) *Investigating the Psychological World: Scientific Method in the Behavioral Sciences*. MIT Press, USA

Harari, N. Y. (2015) *Homo Deus: A Brief History of Tomorrow*. Vintage Books, UK

Hauser, J. R., Liberali, G., Urban, G. L. (2014) 'Website Morphing 2.0: Switching Costs, Partial Exposure, Random Exit, and When to Morph' *Management Science*, 60(6), pp. 1594–1616

Hauser, J. R., Urban, G. L., Liberali, G., Braun, M. (2009) 'Website Morphing' *Marketing Science*, 28(2), pp. 202–223

Hausman, D. M., Welch, B. (2010) 'Debate: To Nudge or Not to Nudge' *Journal of Political Philosophy*, 18(1), pp. 123–136

Hayek, F. A. (1999 [1952]) *The Sensory Order: An Inquiry into the Foundations of Theoretical Psychology*. University of Chicago Press, USA

Heidegger, M. (2010 [1953]) *Being and Time*. State University of New York Press, USA

Heukelom, F. (2012) 'Three Explanations for the Kahneman-Tversky Programme of the 1970s' *The European Journal of the History of Economic Thought*, 19(5), pp. 797–828

Hill, K. (2012) 'How Target Figured Out A Teen Girl Was Pregnant Before Her Father Did' *Forbes*. [Online] [Date accessed: 24/07/2021]: https://www.forbes.com/ sites/kashmirhill/2012/02/16/how-target-figured-out-a-teen-girl-was-pregnant -before-her-father-did/

Hofstadter, D. R. (2000 [1979]) *Gödel, Escher, Bach: an Eternal Golden Braid*. Penguin Books, UK

Jachimowicz, J. M., Duncan, S., Weber, E. U., Johnson, E. J. (2019) 'When and Why Defaults Influence Decisions: A Meta-analysis of Default Effects' *Behavioural Public Policy*, 3(2), pp. 159–186

Johnson, E. J. (2021) 'How Netflix's Choice Engine Drives Its Business' *Behavioral Scientist*. [Online] [Date accessed: 22/10/2021]: https://behavioralscientist.org /how-the-netflix-choice-engine-tries-to-maximize-happiness-per-dollar-spent _ux_ui/

Johnson, E. J., Goldstein, D. G. (2003) 'Do Defaults Saves Lives?' *Science*, 302, pp. 1338–1339

Johnson, E. J., Shu, S. B., Dellaert, B. G. C., Fox, C., Goldstein, D. G., Häubl, G., Larrick, R. P., Payne, J. W., Peters, E., Schkade, D., Wansink, B., Weber, E. U. (2012) 'Beyond Nudges: Tools of a Choice Architecture' *Marketing Letters*, 23, pp. 487–504

Kahneman, D. (2003) 'Maps of Bounded Rationality: Psychology for Behavioral Economics' *The American Economic Review*, 93(5), pp. 1449–1475

Kahneman, D. (2011) *Thinking, Fast and Slow*. Penguin Books, UK

Kahneman, D., Tversky, A. (1972) 'Subjective Probability: A Judgment of Representativeness' *Cognitive Psychology*, 3, pp. 430–454

Kahneman, D., Tversky, A. (1979) 'Prospect Theory: An Analysis of Decision Under Risk' *Econometrica*, 47(2), pp. 263–291

Kahneman, D., Knetsch, J. L., Thaler, R. H. (1991) 'Anomalies: The Endowment Effect, Loss Aversion, and Status Quo Bias' *The Journal of Economic Perspectives*, 5(1), pp. 193–206

Kahneman, D., Sibony, O., Sunstein, C. R. (2021) *Noise: A Flaw in Human Judgement*. William Collins, UK

Kalda, A., Loos, B., Previtero, A., Hackethal, A. (2021) 'Smart(phone) investing? A Within Investor-time Analysis of New Technologies and Trading Behavior' NBER Working Paper 28363. [Online] [Date accessed: 08/02/2021]: www.nber.org/papers/w28363

Kaptein, M., Duplinsky, S. (2013) 'Combining Multiple Influence Strategies to Increase Consumer Compliance' *International Journal of Internet Marketing and Advertising*, 8(1), pp. 32–53

Kaptein, M., Markopoulos, P., de Ruyter, B., Aarts, E. (2015) 'Personalizing Persuasive Technologies: Explicit and Implicit Personalization Using Persuasion Profiles' *International Journal of Human-Computer Studies*, 77, pp. 38–51

Kaptein, M., McFarland, R., Parvinen, P. (2018) 'Automated Adaptive Selling' *European Journal of Marketing*, 52(5–6), pp. 1037–1059

Kelly, C. A., Sharot, T. (2021) 'Individual Differences in Information-seeking' *Nature Communications*, 12, pp. 1–13

Keynes, J. M. (2017 [1936]) *The General Theory of Employment, Interest and Money*. Wordsworth, UK

Knesch, J. L. (1989) 'The Endowment Effect and Evidence for Nonreversible Indifference Curves' *The American Economic Review*, 79(5), pp. 1277–1284

Knetsch, J. L., Sinden, J. A. (1984) 'Willingness to Pay and Compensation Demanded: Experimental Evidence of an Unexpected Disparity in Measures of Value' *The Quarterly Journal of Economics*, 99(3), pp. 507–521

Knetsch, J. L., Sinden, J. A. (1987) 'The Persistence of Evaluation Disparities' *The Quarterly Journal of Economics*, 102(3), pp. 691–695

Kosinski, M., Bachrach, Y., Kohli, P., Stillwell, D., Graepel, T. (2013a) 'Manifestations of User Personality in Website Choice and Behaviour on Online Social Networks' *Machine Learning*, 95, pp. 357–380

Kosinski, M., Stillwell, D., Graepel, T. (2013b) 'Private Traits and Attributes Are Predictable from Digital Records of Human Behavior' *PNAS*, 110(15), pp. 5802–5805

Kramer, A. D. I, Guillory, J. E., Hancock, J. T. (2014) 'Experimental Evidence of Massive-scale Emotional Contagion Through Social Networks' *Proceedings of the National Academy of Sciences of the United States of America*, 111(24), pp. 8788–8790

Kroese, F. M., Marchiori D. R., de Ridder, D. T. (2016) 'Nudging Healthy Food Choices: A Field Experiment at the Train Station' *Journal of Public Health*, 38, pp. 133–137

Krpan, D. Houtsma, N. (2020) 'To Veg or Not to Veg? The Impact of Framing on Vegetarian Food Choice' *Journal of Environmental Psychology*, 67, e. 101391

Krpan, D., Urbanik, M. (2020) 'From Libertarian Paternalism to Liberalism: Behavioural Science and Policy in an Age of New Technology' [Online] [Date accessed: 15/12/2020]: https://psyarxiv.com/5qryd/

Kuhn, T. S. (2012 [1962]) *The Structure of Scientific Revolutions*. University of Chicago Press, USA

Lades, L. K., Delaney, L. (2020) 'Nudge FORGOOD' *Behavioural Public Policy*. https;//doi.org/10.1017/bpp.2019.53

Laffan, K., Sunstein, C. R., Dolan, P. (2021) 'Facing It: Assessing the Immediate Emotional Impacts of Calorie Labelling Using Automatic Facial Coding' *Behavioural Public Policy*. https://doi.org/10.1017/bpp.2021.32

Laibson, D. (1997) 'Golden Eggs and Hyperbolic Discounting' *The Quarterly Journal of Economics*, 112(2), pp. 1937–1996

Langer, M., Oster, D., Speith, T., Hermanns, H., Kästner, L., Schmidt, E., Sesing, A., Baum, K. (2021) 'What do We Want from Explainable Artificial Intelligence (XAI)?: A Stakeholder Perspective on XAI and a Conceptual Model Guiding Interdisciplinary XAI Research' *Artificial Intelligence*, 296, e. 103473

Lanzing, M. (2019) '"Strongly Recommended" Revisiting Decisional Privacy to Judge Hypernudging in Self-Tracking Technologies' *Philosophy and Technology*, 32, pp. 549–568

Lavi, M (2017) 'Evil Nudges' *Journal of Entertainment and Technology Law*, 21(1), pp. 1–93

Lipton, P. (2004) *Inference to the Best Explanation*. Routledge, London

Lipton, Z. C., Steinhardt, J. (2018) 'Troubling Trends in Machine Learning Scholarship' ArXiv. [Online] [Date accessed: 16/09/2021]: https://arxiv.org/abs/1807.03341

Loewenstein, G., Bryce, C., Hagmann, C., Rajpal, S. (2015) 'Warning: You Are About to be Nudged' *Behavioral Science and Policy*, 1(1), pp. 35–42

Lorenz-Spreen, P., Lewandowsky, S., Sunstein, C. R., Hertwig, R. (2020) 'How Behavioural Sciences Can Promote Truth, Autonomy and Democratic Discourse Online' *Nature Human Behaviour*. https://doi.org/10.1038/s41562-020-0889-7

Luckerson, V. (2015) 'Here's How Facebook's News Feed Actually Works' *Time Magazine*. [Online] [Date accessed: 08/03/2021]: https://time.com/collection -post/3950525/facebook-news-feed-algorithm/

Luo, Y., Soman, D., Zhao, J. (2021) 'A Meta-analytic Cognitive Framework of Nudge and Sludge' [Online] [Date accessed: 25/07/2021]: https://psyarxiv.com/dbmu3/ download?format=pdf

Madrian, B. C., Shea, D. F. (2001) 'The Power of Suggestion: Inertia in 401(k) Participation and Savings Behavior' *The Quarterly Journal of Economics*, 116(4), pp. 1149–1187

Marx, K. (2013 [1867]) *Capital*. Wordsworth, UK

Mathur, A., Acar, G., Friedman, M. J., Lucherini, E., Mayer, J., Chetty, M., Narayanan, A. (2019) 'Dark Patterns at Scale: Findings from a Crawl of 11k Shopping Websites' *Proceedings of the ACM Human-Computer Interaction Conference*, 3(81), pp. 1–32

Matthias, A. (2004) 'The Responsibility Gap: Ascribing Responsibility for the Actions of Learning Automata' *Ethics and Information Technology*, 6, pp. 175–183

Matz, S. C., Netzer, O. (2017) 'Using Big Data as a Window into Consumers' Psychology' *Current Opinion in Behavioral Science*, 18, pp. 7–12

Matz, S. C., Kosinki, M., Nave, G., Stillwell, D. J. (2017) 'Psychological Targeting as an Effective Approach to Digital Mass Persuasion' *Proceedings of the National Academy of Sciences of the United States of America*, 114(48), pp. 12714–12719

McCarthy, J. (2007) 'What is Artificial Intelligence?' [Online] [Date accessed: 13/07/2021]: http://jmc.stanford.edu/articles/whatisai/whatisai.pdf

McCarthy, J., Minsky, M. L., Rochester, N., Shannon, C. E. (1955) 'A Proposal for the Dartmouth Summer Research Project on Artificial Intelligence' [Online] [Date accessed: 23/03/2021]: http://www-formal.stanford.edu/jmc/history/dartmouth/dartmouth.html

McKenzie, C. R. M., Liersch, M. J., Finkelstein, S. R. (2006) 'Recommendation Implicit in Policy Defaults' *Psychological Science*, 17(5), pp. 414–420

Mele, C., Spena, T. R., Kaartemo, V., Marzullo, M. L. (2021) 'Smart Nudging: How Cognitive Technologies Enable Choice Architecture for Value Co-creation' *Journal of Business Research*, 129, pp. 949–960

Morozovaite, V. (2021) 'Two Sides of the Digital Advertising Coin: Putting Hypernudging into Perspective' *Market and Competition Law Review*, 5(2), pp. 105–145

Mikolov, T., Chen, K., Corrado, G., Dean, J. (2013) 'Efficient Estimation of Word Representations in Vector Space' ArXiV. [Online] [Date accessed: 04/07/2021]: https://arxiv.org/pdf/1301.3781.pdf

Mill, J. S. (1836) 'On the Definition of Political Economy; and On the Method of Investigation Proper to it' [Online] [Date accessed: 18/01/2022]: https://moehler.org/wp-content/uploads/2021/02/Mill-1844.pdf

Miller, G. A. (1956) 'The Magical Number Seven, Plus or Minus Two: Some Limits on Our Capacity for Processing Information' *Psychological Review*, 101(2), pp. 343–352

Miller, G. A. (2003) 'The Cognitive Revolution: A Historical Perspective' *Trends in Cognitive Science*, 7(3), pp. 141–144

Mills, S. (2020) 'Nudge/Sludge Symmetry: On the Relationship Between Nudge and Sludge and the Resulting Ontological, Normative and Transparency Implications' *Behavioural Public Policy*. https://doi.org/10.1017/bpp.2020.61

Mills, S. (2021) '#DeleteFacebook: From Popular Protest to a New Model of Platform Capitalism?' *New Political Economy*, 26(5), pp. 851–868

Mills, S. (2022) 'Personalized Nudging' *Behavioural Public Policy*, 6(1), pp. 150–159

Mills, S., Whittle, R. (2022) 'Behaviour as a Probability Distribution: Nudging, through the Eye of a Machine' Unpublished Manuscript.

Mills, S., Whittle, R., Brown, G. (2021) 'SpendTech' Unpublished Manuscript.

Mills, S., Sætra, H. S. (2022) 'The Autonomous Choice Architect' *AI and Society*. https://doi.org/10.1007/s00146-022-01486-z

Minsky, M. L., Papert, S. A. (2017 [1969]) *Perceptrons: An Introduction to Computational Geometry*. MIT Press, Cambridge

Mitchell, G. (2005) 'Libertarian Paternalism is an Oxymoron' *Northwestern University Law Review*, 99(3), pp. 1245–1278

Moon, Y. (2010) *Different: Escaping the Competitive Herd*. Crown Business, USA

Morozov, E. (2013) *To Save Everything Click Here: Technology, Solutionism, and the Urge to Fix Problems that Don't Exist*. Allen Lane, UK

Newall, A. (1982) 'The Knowledge Level' *Artificial Intelligence*, 18, pp. 87–127

Newall, A., Simon, H. A. (1972) *Human Problem Solving*. Prentice-Hall, USA

Newall, P. W. S. (2019) 'Dark Nudges in Gambling' *Addiction Research and Theory*, 27(2), pp. 65–67

Noggle, R. (2018) 'Manipulation, Salience, and Nudges' *Bioethics*, 32(3), pp. 164–170

O'Donoghue, T., Rabin, M. (1999a) 'Doing It Now or Later' *The American Economic Review*, 89(1), pp. 103–124

O'Donoghue, T., Rabin, M. (1999b) 'Incentives for Procrastinators' *The Quarterly Journal of Economics* 114(3), pp. 769–816

O'Donoghue, T., Rabin, M. (2001) 'Choice and Procrastination' *The Quarterly Journal of Economics*, 116(1), pp. 121–160

O'Donoghue, T., Rabin, M. (2015) 'Present Bias: Lessons Learned and To Be Learned' *The American Economic Review: Papers and Proceedings*, 105(5), pp. 273–279

Öhman, C., Aggarwal, N. (2020) 'What If Facebook Goes Down? Ethical and Legal Considerations for the Demise of Big Tech' *Internet Policy Review*. https://doi.org/10.14763/2020.3.1488

Oshana, M. A. L. (2002) 'The Misguided Marriage of Responsibility and Autonomy' *The Journal of Ethics*, 6, pp. 261–280

Páez, A. (2019) 'The Pragmatic Turn in Explainable Artificial Intelligence (XAI)' *Minds and Machines*, 29, pp. 441–459

Page, L. C., Castleman, B. L., Meyer, K. (2020) 'Customized nudging to improve FAFSA completion and income verification' *Education Evaluation and Policy Analysis*, 42(1), pp. 3–21

Parkes, D. C., Wellman, M. P. (2015) 'Economic Reasoning and Artificial Intelligence' *Science*, 349(6245), pp. 267–272

Pasquale, F. (2015) *The Black Box Society: The Secret Algorithms That Control Money and Information.* Harvard University Press, USA

Payne, C. R., Niculescu, M., Just, D. R., Kelly, M. P. (2015) 'Shopper Marketing Nutrition Interventions: Social Norms on Grocery Carts Increase Produce Spending Without Increasing Shopper Budgets' *Preventive Medicine Reports*, 2, pp. 287–291

Pedersen, T., Johansen, C. (2020) 'Behavioural Artificial Intelligence: An Agenda for Systematic Empirical Studies of Artificial Inference' *AI and Society*, 35, pp. 519–532

Peer, E., Egelman, S., Harbach, M., Malkin, N., Mathur, A., Frik, A. (2020) 'Nudge Me Right: Personalizing Online Security Nudges to People's Decision-making Styles' *Computers in Human Behavior*, 109, e. 106347

Podsakoff, P. M., MacKenzie, S. B., Podsakoff, N. P. (2016) 'Recommendations for Creating Better Conceptual Definitions in the Organizational, Behavioral, and Social Sciences' *Organizational Research Methods*, 19(2), pp. 159–203

Popper, K. (1959) *The Logic of Scientific Discovery.* Springer, Hutchinson, and Co., USA

Porat, A., Strahilevitz, L. J. (2014) 'Personalized Default Rules and Disclosure with Big Data' *Michigan Law Review*, 112(8), pp. 1417–1478

Possati, L. M. (2020) 'Algorithmic Unconscious: Why Psychoanalysis Helps in Understanding AI' *Palgrave Communications*. https://doi.org/10.1057/s41599-020-0445-0

Possati, L. M. (2021) *The Algorithmic Unconscious: How Psychoanalysis Helps in Understanding AI.* Routledge, UK

Rahwan, I., Cebrian, M., Obradovich, N., Bongard, J., Bonnefon, J., Breazeal, C., Crandall, J. W., Christakis, N. A., Couzin, I. D., Jackson, M. O., Jennings, N. R., Kamar, E., Kloumann, I. M., , Larochelle, H., Lazer, D., McElreath, R., Mislove, A., Parkes, D. C., Pentland, A., Roberts, M. E., Shariff, A., Tenenbaum, J. B., Wellman, M. (2019) 'Machine Behaviour' *Nature*, 568, pp. 477–486

Rauthmann, J. F. (2020) 'A (More) Behavioural Science of Personality in the Age of Multi-Modal Sensing, Big Data, Machine Learning, and Artificial Intelligence' *European Journal of Personality*, 34, pp. 593–598

Rebonato, R. (2014) 'A Critical Assessment of Libertarian Paternalism' *Journal of Consumer Policy*, 37, pp. 357–396

Reinecke, K., Gajos, K. (2014) 'Quantifying Visual Preferences Around the World' IN Proceedings of the 32nd Annual ACM Conference on Human Factors in Computing Systems.

Riceour, P. (2007) *Reflections on the Just*. University of Chicago Press, USA

Richards, T. (1993) *The Imperial Archive: Knowledge and the Fantasy of Empire*. Verso Books, UK

Russell, S. J. (1997) 'Rationality and Intelligence' *Artificial Intelligence*, 94, pp. 57–77

Russell, S. J. (2019) *Human Compatible: AI and the Problem of Control*. Penguin Books, UK

Sadowski, J. (2019) 'When Data Is Capital: Datafication, Accumulation, and Extraction' *Big Data and Society*, 6(1), 1–2, pp. 1–12

Sætra, H. S. (2018) 'Science as a Vocation in the Era of Big Data: The Philosophy of Science behind Big Data and humanity's Continued Part in Science' *Integrative Psychological and Behavioral Science*, 52, pp. 508–522

Sætra, H. S. (2019) 'When nudge comes to shove: liberty and nudging in the era of big data' *Technology in Society*, 59, e. 101130

Sætra, H. S. (2020) 'Privacy as an Aggregate Public Good' *Technology in Society*, 63, e. 101422

Sætra, H. S. (2021a) 'Confound Complexity of Machine Action: A Hobbesian Account of Machine Responsibility' *International Journal of Technoethics*, 12(1), pp. 87–100

Sætra, H. S. (2021b) 'Robotomorphy: Becoming our Creations' *AI and Ethics*. https://doi.org/10.1007/s43681-021-00092-x

Sætra, H. S., Mills, S. (2021) 'Privacy, Surveillance, and Soft Control in the Workplace' Unpublished Manuscript.

Salles, A., Evers, K., Farisco, M. (2020) 'Anthropomorphism in AI' *AJOB Neuroscience*, 11(2), pp. 88–95

Samoili, S., López-Cobo, M., Gómez, E., de Prato, G., Martínez-Plumed, F., Delipetrev, B. (2020) '*AI Watch Defining Artificial Intelligence: Towards an Operational Definition and Taxonomy of Artificial Intelligence*' JRC Technical Reports EUR 30117 EN. [Online] [Date accessed: 25/02/2021]: https://publications.jrc.ec.europa.eu/repository/bitstream/JRC118163/jrc118163_ai_watch._defining_artificial_intelligence_1.pdf

Samuelson, W., Zeckhauser, R. (1988) 'Status Quo Bias in Decision Making' *Journal of Risk and Uncertainty*, 1, pp. 7–59

Sanders, M., Snijders, V., Hallsworth, M. (2018) 'Behavioural Science and Policy: Where Are We Now and Where Are We Going?' *Behavioural Public Policy*, 2(2), pp. 144–167

Schafer, K. (2018) 'A Brief History of Rationality: Reason, Reasonableness, Rationality, and Reasons' *Manuscripto*, 41(4), pp. 501–529

Schöning, C., Matt, C., Hess, T. (2019) 'Personalised Nudging for more Data Disclosure? On the Adaption of Data Usage Policies Format to Cognitive Styles' in Proceeding of the 52nd Hawaii International Conference on System Sciences, pp. 4395–4404

Schultz, P. W., Nolan, J. M., Cialdini, R. B., Goldstein, N. J., Griskevicius, V. (2007) 'The Constructive, Destructive, and Reconstructive Power of Social Norms' *Psychological Science*, 18(5), pp. 429–434

Scott, T. (2017) 'Why The YouTube Algorithm Will Always Be A Mystery' [Online] [Date accessed: 06/07/2021]: https://www.youtube.com/watch?v =BSpAWkQLlgM

Seaver, N. (2017) 'Algorithms as Culture: Some Tactics for the Ethnography of Algorithmic Systems' *Big Data and Society*. https://doi.org/10.1177 /2053951717738104

Selinger, E., Whyte, K. P. (2010) 'Competence and Trust in Choice Architecture' *Knowledge, Technology and Policy*, 23, pp. 461–482

Service, O., Hallsworth, M., Halpern, D., Algate, F., Gallagher, R., Nguyen, S., Ruda, S., Sanders, M., Pelenur, M., Gyani, A., Harper, H., Reinhard, J., Kirkman, E. (2015) 'EAST: Four Simple Ways to Apply Behavioural Insights' *BIT*. [Online] [Date accessed: 24/07/2021]: http://www.behaviouralinsights.co.uk/wp-content/ uploads/2015/07/BIT-Publication-EAST_FA_WEB.pdf

Sharif, A., Moorlock, G. (2018) 'Influencing Relatives to Respect Donor Autonomy: Should We Nudge Families to Consent to Organ Donation?' *Bioethics*, 32(3), pp. 155–163

Sharot, T., Sunstein, C. R. (2020) 'How People Decide What They Want to Know' *Nature Human Behaviour*, 4, pp. 14–19

Shelton, K. (2017) 'The Value of Search Results' *Forbes*. [Online] [Date accessed: 25/07/2021]: https://www.fca.org.uk/insight/robo-advice-%E2%80%93-will -consumers-get-programme

Skinner, B. F. (1953) *Science and Human Behavior*. Macmillan, USA

Skinner, B. F. (1971) *Beyond Freedom and Dignity*. Knopf, USA

Skinner, B. F. (1976 [1974]) *About Behaviorism*. Vintage, USA

Skinner, B. F. (1984) 'An Operant Analysis of Problem Solving' *Behavioral and Brain Sciences*, 7, pp. 583–613

Silver, D., Singh, S., Precup, D., Sutton, R. S. (2021) 'Reward Is Enough' *Artificial Intelligence*, 209, e. 103535

Simkulet, W. (2018) 'Informed Consent and Nudging' *Bioethics*, 33(1), pp. 169–184

Simon, H. A. (1955) 'A Behavioral Model of Rational Choice' *The Quarterly Journal of Economics*, 69(1), pp. 99–118

Simon, H. A. (1956) 'Rational Choice and the Structure of the Environment' *Psychological Review*, 63(2), pp. 129–138

Simon, H. A. (1977) 'What Computers Mean for Man and Society' *Science*, 195, pp. 1186–1191

Simon, H. A. (1981) 'Information-Processing Models of Cognition' *Journal of the American Society for Information Science*, 32(5), pp. 364–377

Simon, H. A. (1994 [1969]) *The Sciences of the Artificial*. MIT Press, USA

Smith, J., de Villiers-Botha, T. (2021) 'Hey, Google, leave those kids alone: against hypernudging children in the age of big data' *AI and Society*. DOI: 10.1007/s00146-021-01314-w

Smith, A., Harvey, J., Goulding, J., Smith, G., Sparks, L. (2020) 'Exogenous Cognition and Cognitive State Theory: The Plexus of Consumer Analytics and Decision-making' *Marketing Theory*. https://doi.org/10.1177/1470593120964947

Soman, D. (2020) 'Sludge: A Very Short Introduction' [Online] [Date accessed: 25/07/2021]:   https://www.rotman.utoronto.ca/-/media/Files/Programs-and -Areas/BEAR/White-Papers/BEARxBIOrg-Sludge-Introduction.pdf?la=en &hash=DCB98795CB485977A04DDB27EFD800C3DA40220E

Soman, D., Zhao, M. (2011) 'The Fewer the Better: Number of Goals and Savings Behavior' *Journal of Marketing Research*, 48(6), pp. 944–957

Soman, D., Xu, J., Cheema, A. (2010) 'Decision Points: A Theory Emerges' *Rotman Magazine*, Winter, pp. 64–68

Sparkman, G., Walton, G. M. (2017) 'Dynamic Norms Promote Sustainable Behavior, Even if It Is Counternormative' *Psychological Science*, 28(11), pp. 1663–1674

Spencer, D. (2018) 'Work in and Beyond the Second Machine Age: The Politics of Production and Digital Technologies' *Work, Employment and Society*, 31(1), pp. 142–152

Srnicek, N. (2016) *Platform Capitalism*. Polity Books, UK

Steffel, M., Williams, E. F., Pogacar, R. (2016) 'Ethically Deployed Defaults: Transparency and Consumer Protection Through Disclosure and Preference Articulation' *Journal of Marketing Research*, 53, pp. 865–880

Sternberg, R. J. (1999) 'The Theory of Successful Intelligence' *Review of General Psychology*, 3(4), pp. 292–316

Sternthal, B., Dholakia, R., Leavitt, C. (1978) 'The Persuasive Effect of Source Credibility: Tests of Cognitive Response' *Journal of Consumer Research*, 4(4), pp. 252–260

Sunstein, C. R. (1996), 'Social Norms and Social Roles' *Columbia Law Review*, 96(4), pp. 903–968.

Sunstein, C. R. (2012) 'Impersonal Default Rules vs. Active Choices vs. Personalized Default Rules: A Triptych' SSRN. [Online] [Date accessed: 20/01/2020]: https://ssrn.com/abstract=2171343

Sunstein, C. R. (2013) 'The Storrs Lectures: Behavioral Economics and Paternalism' *The Yale Law Journal*, 122, pp. 1826–1899

Sunstein, C. R. (2014) *Why Nudge? The Politics of Libertarian Paternalism*. Yale University Press, USA

Sunstein, C. R. (2017) '*Misconceptions About Nudges*' SSRN. [Online] [Date accessed]: 15/04/2020]: https://ssrn.com/abstract=3033101

Sunstein, C. R. (2019a) 'Sludge Audits' *Behavioural Public Policy*. https://doi.org/10.1017/bpp.2019.32

Sunstein, C. R. (2019b) 'Algorithms, Correcting Biases' *Social Research: An International Quarterly*, 86(2), pp. 499–511

Sunstein, C. R. (2019c) 'Ruining popcorn? The welfare effects of information' *Journal of Risk and Uncertainty*, 58, pp. 121–142

Sunstein, C. R. (2019d) 'Sludge Audits' *Behavioural Public Policy*, https://doi.org/10.1017/bpp.2019.32

Sunstein, C. R. (2020a) *Too Much Information: Understanding What You Don't Want to Know*. MIT Press, USA

Sunstein, C. R. (2020b) 'Behavioral Welfare Economics' *Journal of Benefit-Cost Analysis*, 11(2), pp. 196–220

Sunstein, C. R. (2021a) 'Hayekian behavioral economics' *Behavioural Public Policy*. https://doi.org/10.1017/bpp.2021.3

Sunstein, C. R. (2021b) 'The distributional effects of nudges' *Nature Human Behaviour*. https://doi.org/10.1038/s41562-021-01236-z

Susser, D. (2019) 'Invisible Influence: Artificial Intelligence and the Ethics of Adaptive Choice Architectures' in AIES'19. https://doi.org10.1145/3306618.3314286

Tegmark, M. (2017) *Life 3.0: Being Human in the Age of Artificial Intelligence*. Penguin Books, UK

Thaler, R. H. (1980) 'Toward a Positive Theory of Consumer Choice' *Journal of Economic Behavior and Organization*, 1, pp. 39–60

Thaler, R. H. (1999) 'Mental Accounting Matters' *Journal of Behavioral Decision Making*, 12(3), pp. 183–206

Thaler, R. H. (2008) 'Mental Accounting and Consumer Choice' *Marketing Science*, 27(1), pp. 15–25

Thaler, R. H. (2015) *Misbehaving: The Making of Behavioural Economics*. Penguin Books, UK

Thaler, R. H. (2018) 'Nudge, not Sludge' *Science*, 361(6401), pp. 431–432

Thaler, R. H. (2021) 'What's Next for Nudging and Choice Architecture?' *Organizational Behavior and Human Decision Processes*, 163, pp. 4–5

Thaler, R. H., Sunstein, C. R. (2003) 'Libertarian Paternalism' *The American Economic Review*, 93(2), pp. 175–179

Thaler, R. H., Sunstein, C. R. (2008) *Nudge: Improving Decisions about Health. Wealth and Happiness*' Penguin Books, UK

Thaler, R. H., Tucker, W. (2013) 'Smarter Information, Smarter Consumers' *Harvard Business Review*, 91(1–2), pp. 44–54

Thaler, R. H., Sunstein, C. R., Balz, J. (2012) 'Choice Architecture' in Shafir, E. (eds.) *The Behavioral Foundations of Public Policy*, Princeton University Press, USA

Thunström, L. (2019) 'Welfare Effects of Nudges: The Emotional Tax of Calorie Menu Labeling' *Judgment and Decision Making*, 14(1), pp. 11–25

Thunström, L., Gilbert, B., Jones-Ritten, C (2018) 'Nudges that Hurt Those Already Hurting: Distributional and Unintended Effects of Salience Nudges' *Journal of Economic Behavior and Organization*, 153, pp. 267–282

Thunström, L., Nordström, J., Shogren, J. F., Ehmke, M., van't Veld, K. (2016) 'Strategic Self-ignorance' *Journal of Risk and Uncertainty*, 52(2), pp. 117–136

Turkle, S. (1988) 'Artificial Intelligence and Psychoanalysis: A New Alliance' *Dædalus*, 117(1), pp. 241–268

Turkle, S. (2004 [1984]) *The Second Self: Computers and the Human Spirit*. MIT Press, USA

Tversky, A., Kahneman, D. (1973) 'Availability: A Heuristic for Judging Frequency and Probability' *Cognitive Psychology*, 5, pp. 207–232

Tversky, A., Kahneman, D. (1974) 'Judgment under Uncertainty: Heuristics and Biases' *Science*, 185(4157), pp. 1124–1131

Tversky, A., Kahneman, D. (1992) 'Advances in Prospect Theory: Cumulative Representation of Uncertainty' *Journal of Risk and Uncertainty*, 5, pp. 297–323 van den Hoven, M. (2020) 'Nudging for Others' Sake: An Ethical Analysis of the Legitimacy of Nudging Healthcare Workers to Accept Influenza Immunization' *Bioethics*, 35(2), pp. 143–150

van der Linden, S., Maibach, E., Cook, J., Leiserowitz, A., Lewandowsky, S. (2017) 'Inoculating Against Misinformation' *Science*, 358(6367), pp. 1141–1142

van Fraassen, B. C. (1989) *Laws and Symmetry*. Oxford University Press, UK

Varian, H. (2009) 'Discussion of "Website Morphing"' *Marketing Science*, 28(2), pp. 224

Veblen, T. (2012 [1899]) *The Theory of the Leisure Class*. Renaissance Classics, USA

Veritasium (2019) 'My Video Went Viral. Here's Why' [Online] [Date accessed: 06/07/2021]: https://www.youtube.com/watch?v=fHsa9DqmId8

Viljoen, S. (2020) 'Democratic Data: A Relational Theory for Data Governance' *SSRN*. [Online] [Date accessed: 11/12/2020]: https://ssrn.com/abstract=3627562

Villanova, D., Bodapati, A. V., Puccinelli, N. M., Tsiros, M., Goodstein, R. C., Kushwaha, T., Suri, R., Ho, H., Brandon, R., Hatfield, C. (2021) 'Retailer Marketing Communications in the Digital Age: Getting the Right Message to the Right Shopper at the Right Time' *Journal of Retailing*, 97(1), pp. 116–132

Villiappan, N., Dai, N., Steinberg, E., He, J., Rogers, K., Ramachandran, V., Xu, P., Shojaeizadeh, M., Guo, L., Kohlhoff, K., Navalpakkam, V. (2020) 'Accelerating Eye Movement Research via Accurate and Affordable Smartphone Eye Tracking' *Nature Communications*, 11(4553), pp. 1–12

Vincent, J. (2016) 'Twitter Taught Microsoft's AI Chatbot to be a Racist Asshole in Less than a Day' *The Verge*. [Online] [Date accessed: 04/07/2021]: https://www.theverge.com/2016/3/24/11297050/tay-microsoft-chatbot-racist

Vincent, J. (2020) 'Facebook Is Now Using AI to Sort Content for Quicker Moderation' *The Verge*. [Online] [Date accessed: 22/03/2021]: https://www.theverge.com/2020/11/13/21562596/facebook-ai-moderation

von Neumann, J. (2000 [1958]) *The Computer and the Brain*. Yale University Press, USA

von Neumann, J., Morgenstern, O. (1944) *Theory of Games and Economic Behavior*. Princeton University Press, USA

Wall Street Journal (2021) 'Inside TikTok's Algorithm: A WSJ Video Investigation' [Online] [Date accessed: 28/10/2021]: https://www.wsj.com/articles/tiktok-algo-rithm-video-investigation-11626877477

Waters, M. (2021) 'Millennials, What Will It Take for You to Buy Life Insurance?' *The Atlantic*. [Online] [Date accessed: 01/08/2021]: https://www.theatlantic.com/technology/archive/2021/07/great-life-insurance-rebrand/619603/?utm_campaign=the-atlantic&utm_medium=social&utm_term=2021-07-30T11%3A00%3A57&utm_content=edit-promo&utm_source=twitter

Watson, D. (2019) 'The Rhetoric and Reality of Anthropomorphism in Artificial Intelligence' *Minds and Machines*, 29, pp. 417–440

Watson, J. B. (1913) 'Psychology as the Behaviorist Views It' *Psychological Review*, 20(2), pp. 158–177

Weinmann, M., Schneider, C., vom Brocke, J. (2016) 'Digital Nudging' *SSRN*. [Online] [Date accessed: 30/01/2020]: https://papers.ssrn.com/sol3/papers.cfm?abstract_id=2708250

Wiggers, K. (2020) 'Amazon's AI Generates Images of Clothing to Match Text Queries' *Venture Beat*. [Online] [Date accessed: 24/03/2021]: https://venturebeat.com/2020/03/02/amazons-ai-generates-images-of-clothing-to-match-text-queries/

Willis, L. E. (2020) 'Deception by Design' *Harvard Journal of Law and Technology*, 34(1), pp. 115–190

Yeung, K. (2017) '"Hypernudge": Big Data as a mode of regulation by design' *Information, Communication and Society*, 20(1), pp. 118–136

Zarouali, B., Poels, K., Ponnet, K., Walrave, M. (2018) '"Everything Under Control?": Privacy Control Salience Influences both Critical Processing and Perceived Persuasiveness of Targeted Advertising Among Adolescents' *Journal of Psychosocial Research on Cyberspace*, 12(1), pp. 1–5

Zerilli, J., Knott, A., Maclaurin, J., Gavaghan, C. (2019) 'Algorithmic Decision-Making and the Control Problem' *Minds and Machines*, 29, pp. 555–578

Zittrain, J. (2014) 'Engineering an Election' *Harvard Law Review*, 127, pp. 335–341

Zuboff, S. (1988) *In the Age of the Smart Machine: The Future of Work and Power.* Basic Books, USA

Zuboff, S. (2015) 'Big Other: Surveillance Capitalism and the Prospects of an Information Civilization' *Journal of Information Technology*, 30, pp. 75–89

Zuboff, S. (2019) *The Age of Surveillance Capitalism: The Fight for A Human Future At the New Frontier of Power.* Profile Books, UK

Index

Notes: The page preference with letter n represents note numbers.

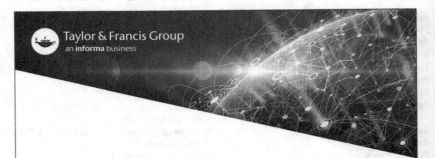

Printed in the United States
by Baker & Taylor Publisher Services

Printed in the United States
by Baker & Taylor Publisher Services